CHEMICAL INTOLERANCE

Physiological Causes and Effects and Treatment Modalities

Robert W. Gardner, Ph.D.
Brigham Young University
Provo, Utah

CRC Press
Boca Raton Ann Arbor London Tokyo

Library of Congress Cataloging-in-Publication Data

Gardner, Robert W., Ph.D.
Chemical intolerance : physiological causes and effects and treatment modalities / Robert W. Gardner.
p. cm.
Includes bibliographical references and index.
ISBN 0-8493-8926-7
1. Immunotoxicology. 2. Chemicals—Health aspects. 3. Environmentally induced diseases. 4. Food allergies. I. Title.
[DNLM: 1. Hypersensitivity—etiology. 2. Environmental Pollutants—adverse effects. 3. Environmental Exposure—adverse effects. WD 300 G228c 1993]
RC582.17.G37 1993
615.9′.02—dc20
DNLM/DLC
for Library of Congress 93-16010
CIP

Direct all inquiries to CRC Press, Inc., 2000 Corporate Blvd., N.W., Boca Raton, Florida 33431.

International Standard Book Number 0-8493-8926-7
Library of Congress Card Number 93-16010
Printed in the United States of America 1 2 3 4 5 6 7 8 9 0
Printed on acid-free paper

Preface

Causes and effective treatments of inflammatory and allergic reactions are still elusive, even with the availability of modern medical technology. An increasing awareness of the toxicological and pharmacological effects of chemicals found as normal constituents of foods and pollens, as well as food additives and air pollutants, has resulted in valuable findings relative to physiological responses and affected biochemical pathways. Phenolic (aromatic) compounds particularly have been the focus of attention, both for their adverse pharmacological effects and for their therapeutic value when used at proper dosage levels. Progressive knowledge of the factors affecting the synthesis of eicosanoids (prostaglandins, thromboxanes, and leukotrienes) and their biological effects has now made it possible to identify chemicals which activate the synthesis of eicosanoids. Chemicals are also being identified which inhibit the cyclooxygenase and lipoxygenase pathway production of the eicosanoids. This has important therapeutic value in treating a large array of human disease problems associated with inflammatory and allergic reactions.

The first chapter of the book includes a study of chemicals which have been identified as causing adverse biological effects and the biochemical reasons for those effects. Chapters 2 through 13 are devoted to specific clinical effects attributable to the chemical activators discussed in Chapter 1. Chemicals which have proven to be of pharmacological value in treating the disorders described in Chapters 2 through 13 are discussed in Chapter 14, which also includes treatment modalities relative to dosage levels required to be both safe and effective. In the final chapter, the author shares his findings of 20 years of research motivated by his own total food intolerance and multiple chemical sensitivities. This includes a listing of the few "key" chemicals which he found arrested reactions to all the others. The Appendix contains results of a clinical study verifying theories proposed in the book.

This publication is written for scientists, clinicians, and the pharmaceutical industry to stimulate additional research and to provide relief to millions who are stressed with chemical sensitivities. Scientific references provided should be a catalyst for others to verify and expand on these discoveries. With the urging of medical associates who are using these findings, the author has endeavored to assemble the findings of numerous scientists and to combine them with his own findings in an effort to offer a biochemical basis for the multi-faceted clinical manifestations of chemical sensitivities.

The Author

Robert W. Gardner, Ph.D., is Professor Emeritus of Animal Science at Brigham Young University, Provo, Utah. Dr. Gardner obtained a B.Sc. in Animal Science at Utah State University in 1959 and his M.Sc. and Ph.D. at Cornell University in Animal and Human Nutrition in 1962 and 1964, respectively, with additional study in biochemistry and physiology. His research in animal nutrition at Brigham Young University was interrupted when he discovered that he had total food intolerance and that a cure was not available from the medical profession. He subsequently identified the phenolic compounds which are found naturally in foods, pollens, etc., and are the principal cause of food intolerance. These compounds were found to have pharmaceutical effects, rather than immunological effects. Dr. Gardner then determined that, when taken in correct dosages, certain phenolics suppress formation of inflammatory eicosanoids (prostaglandins, thromboxanes, and leukotrienes). He has presented the results of his research to numerous medical groups, and after one such presentation he was asked to serve as an honorary member of the Neuroallergy Committee of the American College of Allergists.

The International Society for the Study of Biochemical Intolerance was organized in 1984 by a group of medical doctors who have successfully applied Dr. Gardner's findings in their own clinical practices. Members of the Society met annually to share Dr. Gardner's findings from his 20 years of continuous testing and literature research.

Dr. Gardner was a member of the University Research Committee and the Faculty Advisory Council of Brigham Young University; he served for many years as guest editor of the *Journal of Applied Physiology*; and he has been invited to present a variety of papers throughout his career as a consultant, teaching professor, and research scientist.

Table of Contents

An Introduction and Brief History

Persistent diarrhea, flatulence, and associated illness caused me to seek a diagnosis by medical doctors over 20 years ago (1965 to 1974). Their prescriptions only accentuated the problem. Cromolyn sodium (Intal®) in tablet form (dose level) triggered more diarrhea, a dry mouth, and other side effects. Corticosteroids caused stressful side effects, including retention of fluids. No pathogens (including yeast) were found in stool cultures; however, the pH of the stools was 4.0. A gastroenterologist was eventually consulted. He prescribed a digestive enzyme preparation. Results were negative. A small-bowel biopsy followed, which showed normal villi. No signs of ulceration were found in the stomach. A proctoscopic examination ruled out colitis as the pathological factor. Subsequently, an exploratory operation revealed an inflamed ("blushing") appearance in the last 2 feet of the small intestine. A biopsy from that section revealed nothing pathological, only vascular engorgement. The gastroenterologist attributed causation to motor nerve derangement in the gut, but had no explanation for the factor(s) responsible for the abnormal peristaltic action. A codeine/Amytal® capsule combination was prescribed to slow intestinal motility. Food allergies were then considered as a possible cause. An allergist was eventually consulted. He verified sensitivities to a large number of foods by using intradermal tests. However, classical immunological reactions seemed unlikely since an immunoglobulin E (IgE) test value was one fifth of normal. A rigid rotation diet was pursued with no remission of symptoms. The allergist himself had chemical sensitivities and taught me the method of preparing and using sublingual neutralizing doses of food extracts. He helped me acquire 86 different food extracts, and I then made 1:10 dilutions until I found dose levels which would normalize my pulse and simultaneously relieve symptoms. This was a very frustrating procedure since reactions to every food needed to be neutralized after each meal and the neutralizing dilutions were changing.

This allergist also informed me that I was sensitive to phenol and salicylates (such as aspirin). Was I possibly reacting to phenolic compounds found naturally in foodstuffs as well as in food additives? The allergist challenged me to use my doctoral education in nutrition, biochemistry, and physiology to find the underlying cause of these sensitivities. I began by searching in the university library for possible sources and effects of phenolic compounds. This resulted in finding such publications as *Plant Phenolics, Pharmacology of Plant Phenolics, Organic Constituents of Higher Plants, Toxicants Occurring Naturally in Foods, Toxic Constituents of Plant Foodstuffs, POLLEN: Biology, Biochemistry, Management,* and Singleton's publications on phenolic substances of plant origin common in foods which are naturally occurring food toxicants. With those clues I acquired pure phenolic compounds, made dilutions, tested for sensitivities, and administered neutralizing doses via sublingual challenge. I found that I reacted to each phenolic compound I tested (i.e., caffeic acid, gallic acid, quercetin, benzyl isothiocyanate, etc.). Initial testing was with microgram amounts of each compound. Next came sublingual test challenges with food extracts to determine which phenolic compounds might arrest reactions to specific foods. A dot chart was created which matched foods with phenolic compounds capable of neutralizing reactions to those foods. Interestingly, a few of those phenolics were effective in curbing reactions to a broad spectrum of foods, even though they were not in the same botanical food families. The reason will become clear when one reads the text of the book, particularly Chapters 14 and 15. Joseph J. McGovern, M.D., of Oakland, California, learned of my findings and asked me to share my observations so that he might test this concept on patients in his allergy practice whom he had found did not respond to current medical treatments for allergy. He, in collaboration with L. D. Brenneman, M.D., Ph.D., of the University of California, San Francisco, conducted a double-blind study using 100 patients with food allergy and phenol intolerance. Laboratory analysis revealed that these patients had elevated circulating immune complexes, elevated prostaglandin $F_{2\alpha}$, decreased IgE (RAST [radioallergosorbent test] usually negative), decreased IgA, and elevated IgM. The results of the study were reported at the 37th Annual Congress of the American College of Allergists in 1981 and are included in the appendix of this book. After we made that presentation, I was invited by Dr. Leonard Girsh, Chairman of the Neuroallergy Committee of the College of Allergists, to serve as an honorary member of that committee. This was an impetus to search for chemical factors which affect the central nervous system, to identify biochemical mechanisms involved, and to determine the most effective chemicals needed in therapy. My own experience with central nervous disorders has been incentive enough, since during the course of this sickness I have experienced cases of convulsive seizures while sleeping; have been plagued with deep depression on many occasions; have had semiblackouts, vertigo, chronic fatigue, impaired mental function associated with lethargy and fatigue, drowsiness, withdrawal syndrome, memory lapses; and have become irritable and filled with tension when severely stressed by reactions. Happily, the majority of those symptoms have been brought under control as I have found the chemicals to modulate responses by neurotransmitters. An additional event which should be recorded in this neurological history is the delirium I experienced when given an injection of codeine at the emergency room of a hospital to reduce the pain of a strep throat. This occurred prior to finding the sensitivities to phenolics and, in retrospect, was a reaction to yet another phenolic compound.

The findings of Drs. McGovern and Brenneman alerted me to the possible role of prostaglandins in food, pollen, and general chemical sensitivities. Known clinical effects of prostaglandins duplicated many of the symptoms I had been experiencing. Knowledge of the biological effects of chemicals which either activate or curb the synthesis of prostaglandins, thromboxanes, and leukotrienes has greatly expanded since the late 1970s. The specific roles of phenolic (aromatic) compounds in these processes have been reported independently by scientists in many countries. Now there was a rational biochemical reason for the use of phenolic compounds at neutralizing dose levels. These are pharmacological effects, and dosage levels are critical in order to suppress reactions in one biochemical pathway without simultaneously shunting reactions into other pathways.

A medical doctor urged me to take megadoses of vitamins and essential minerals. In theory this was "to build up the body and thus help detoxify body systems". In fact, I reacted to several of the vitamins (niacin, riboflavin, thiamine, pyridoxine, vitamin B_{12}, biotin, vitamin A, vitamin D, vitamin E, and vitamin K — naphthoquinones). These are also aromatic compounds and may have toxic effects by disturbing metabolic pathways. Calcium, magnesium, sodium chloride, and potassium chloride also caused disturbances.

Chapter 15 is a continuation of this history, since it includes my findings of which compounds were most effective in controlling adverse reactions. Many of these findings have been shared with collaborating medical doctors who have in turn found them effective in treating patients with allergies, particularly patients who have not responded to conventional injection therapy. They have encouraged me to write this history and share these discoveries so that others may be spared the stress of chemical sensitivities associated with foods, food additives, and inhaled materials.

1

Chemical Activators of Allergic and Inflammatory Reactions

I. INTRODUCTION

Proteins traditionally have been implicated as the chemical factors activating allergic reactions.[1] Immunoglobulins, particularly immunoglobulin E (IgE) antibodies, have been designated as the causative factors.[1,2] Hyperimmunoglobulin E syndrome is a recent designation for patients with a triad of severe eczematoid dermatitis, recurrent infections, and very high levels of IgE.[3] Antigens are purportedly eliminated by antibody, complement, and phagocytes, with a likelihood that defective function of one of these, primary (genetic) or secondary, could lead to defective elimination.[4]

Simple chemical compounds and certain drugs are considered to be haptens.[5] Chemical bonding between such a chemical (hapten) and primary amino or sulfhydryl groups of proteins is presumed to create a compound which results in immunogenicity and allergy.[5] Another consideration is that certain chemicals in food additives — pollens, preservatives, antioxidants, dyes, etc. — may activate an inflammatory effect, a pharmacological effect, nonimmunologic hypersensitivity with "allergy-like" symptoms, or pseudo-allergy.[6]

II. INFLAMMATORY AGENTS

Eicosanoids (prostaglandins, thromboxanes, and leukotrienes) as well as histamine and serotonin have been identified with hypersensitivity or inflammatory reactions.[7] For instance, an antigen combination with reaginic or homocytotropic antibody on the surfaces of mast cells or basophils results in the release of these mediators (Figure 1.1).[7] The antibody is usually IgE.[7] Eicosanoids may also be generated by vascular and other tissues by nonimmunologic stimuli as well.[8]

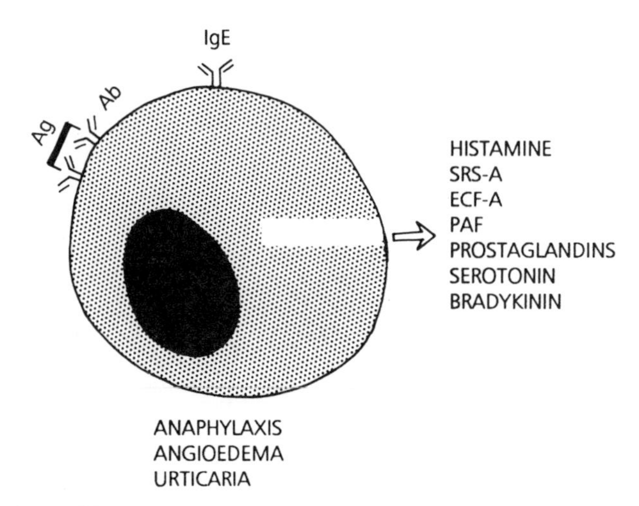

Figure 1.1 Type I hypersensitivity reaction. Immediate hypersensitivity. SRS-A = slow-reacting substance of anaphylaxis, ECF = eosinophil chemotatic factor A, PAF = platelet activating factor, Ag = antigen, Ab = antibody. (From Chandra, R. K. and Shah, A., in *Food Intolerance,* Chandra, R. K., Ed., Elsevier, New York, 1984, 63. With permission.)

A. Phenolic Compound Activation of the Inflammatory Response

Phenolic compounds present in plant foodstuffs and pollens (Figure 1.2), food additives (Figure 1.3), tobacco smoke, perfumes, air pollution, etc. are also important activators in the synthesis of eicosanoids, particularly in chemically sensitive patients.[9] These are considered to be nonimmunologic mechanisms associated with symptoms described as "allergic". Patients presenting with symptoms of allergies, but with normal or subnormal levels of IgE, are very possibly chemically sensitive patients.[9] Relative frequency figures of immunologic vs. nonimmunologic allergic reactions could not be found. However, with inflammatory mediators common to both, treatment medications may be developed which prevent the synthesis, release, or blockage of receptor activation sites of such mediators.

Pharmacological and toxicological effects of plant phenolics in foods and pollens have been recognized for a number of years.[10-14] Many of these natural chemical toxins protect plants against fungi, insects, and animal predators.[13,14] Ames et al.[13] estimate that Americans eat about 1.5 g of natural pesticides per person per day, which is about 10,000 times more than they eat of synthetic pesticide residues. Those figures were used to conclude that dietary pesticides are 99.99% natural.[13]

Other dietary sources of phenolics include food additives used to artificially color, flavor, preserve, or sweeten foods. Examples are tartrazine (FD&C yellow #5), FD&C blue #2, erythrosine (FD&C red #3), vanillin, eugenol, propyl gallate, propylparaben, butylated hydroxyanisole, butylated hydroxytoluene, and aspartame (Figure 1.3). Lockey[15] was one of the

	Monohydroxy	Dihydroxy	Trihydroxy
Benzoic acids	*p*-hydroxybenzoic acid	R=H, protocatechuic acid R=CH_3, vanillic acid	R= H, gallic acid R= CH_3, syringic acid
Cinnamic acids	*p*-coumaric acid	R=H, caffeic acid R=CH_3, ferulic acid	sinapic acid
Flavonols	kaempferol	quercetin	myricetin
Anthocyanidins		cyanidin	R=H, delphinidin R=CH_3, malvidin
Leucoanthocyanidin (flavan-3,4-diols)		leucocyanidin	leucodelphinidin

Figure 1.2 Phenolic compounds commonly found in plant foods and pollens. (From Ribereau-Gayon, P., in *Plant Phenolics*, Heywood, V. H., Ed., Hafner Publishing, New York, 1972, 10. With permission.)

first medical doctors to report hypersensitivity to tartrazine, other dyes, and food additives. Responses ranging from anaphylaxis to urticaria, headaches, and asthma attacks have been reported in patients sensitive to tartrazine.[15-17] The ingestion of aspartame induced significant increases in catecholamine neurotransmitters (particularly norepinephrine) in several brain regions of mice.[18] This could account for the report of the triggering of migraine attacks in a 31-year-old woman whenever she drank a soft drink containing this artificial sweetener or took aspartame tablets.[19]

Plant flavonoids (Figure 1.2), a novel group of phenolic compounds, have recently been the focus of intensive investigations.[20,21] Biochemical, pharmacological, and structure/activity relationships of flavonoids have been defined,[20] as well as cellular and medicinal properties.[21] An important association has been observed between several phenolic compounds (quercetin, esculetin, phenol, tangeretin, naringenin, malvidin, gallic acid, chlorogenic acid, serotonin, etc.) and the production of eicosanoids.[20-22] Numerous phenolic compounds potentiate the toxicity of epinephrine and norepinephrine (catecholamines) by interfering with their oxidation (inactivation).[11,22-25] These catecholamines and other neurotransmitters stimulate the synthesis and release of eicosanoids.[24,25] A proposed mechanism for such activation is through the adenylate cyclase enzyme via β-1 subtype receptors to convert adenosine triphosphate (ATP) to cyclic adenosine

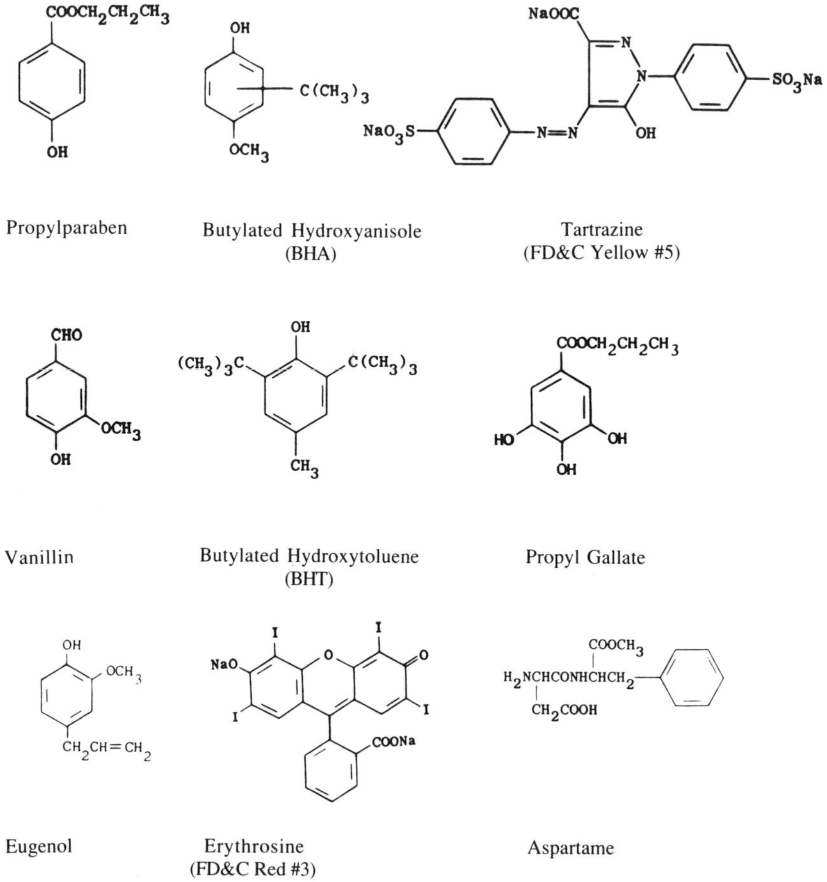

Propylparaben

Butylated Hydroxyanisole (BHA)

Tartrazine (FD&C Yellow #5)

Vanillin

Butylated Hydroxytoluene (BHT)

Propyl Gallate

Eugenol

Erythrosine (FD&C Red #3)

Aspartame

Figure 1.3 Examples of phenolic compounds used as food additives.

monophosphate (cAMP; Figure 1.4).[26] In turn, cAMP activates the liberation of arachidonic acid from triglycerides via an increase in phosphatidylcholine synthesis.[27] Another initiating signal is the activation of phospholipase C and/or phospholipase A_2 (PLA_2), which may be activated by intracellular messengers such as cAMP, protein kinase C, Ca^{2+}, or diacylglycerol (Figures 1.5 and 1.6).[24,29-32] Activation of PLA_2 can also be mediated by acetylcholine, but this effect is about one third that of norepinephrine.[32] Muscarinic receptors are activated in this process, resulting in the liberation of prostaglandins.[32]

Positioning of fatty acids in triglycerides and the specificity of action of the lipase enzymes are illustrated by Moore[33] in Figure 1.5.

An unusual source of PLA_2 is bee venom.[34,35] This is considered to be the major allergen in bee venom. An IgE antibody response to PLA_2 is cited as the cause of more than 90% of patients with bee venom anaphylaxis.[34] However, the role of PLA_2 in releasing arachidonic acid and subsequent

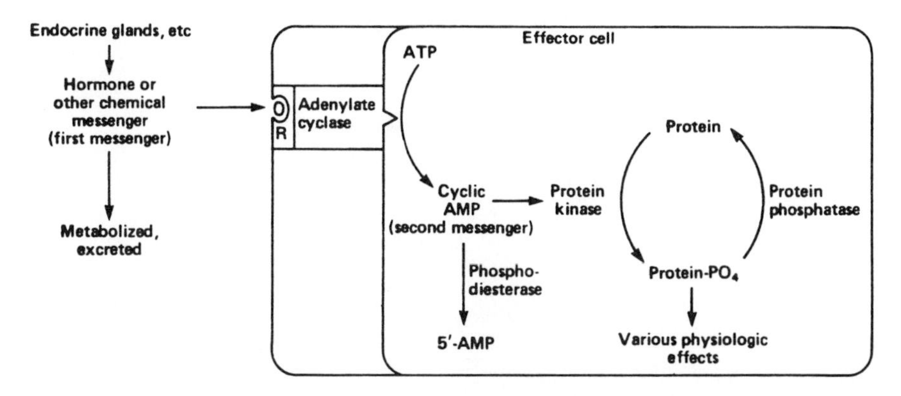

Figure 1.4 The role of cAMP as a second messenger. R = receptor for first messenger. (From Ganong, W. F., *Review of Medical Physiology*, 9th ed., 1979, Lange Medical Publications, Los Altos, California. Modified from Greengard, P., Phosphorylated proteins as physiological effectors, *Science*, 199, 146, 1978. With permission.)

production of eicosanoids should be considered in addition to the IgE antibody response.

The concentration of PLA_2 in individual European bee venoms varied between 1.8% and 27.4% (w/w) in a study reported by Schumacher et al.[35] Africanized bees contained less venom but more PLA_2 than did the European bees; consequently, risks from stings varied as a factor of PLA_2 in the venom.

cAMP concentrations may remain abnormally high due to the presence of caffeine or other methyl xanthines, primarily by inhibition of cAMP phosphodiesterase (Figure 1.4).[26,36] This in turn activates release of arachidonic acid with subsequent formation of eicosanoids. Ascorbic acid (vitamin C) in large doses may also possibly cause adverse effects in some individuals due to the role of the vitamin as a nutrient involved in both the formation and protection from oxidation of norepinephrine and epinephrine.[37]

B. Stress

Stress is possibly an activating factor in the synthesis of eicosanoids inasmuch as stress normally leads to acidosis.[38] Norepinephrine, one of the most effective buffers in the body, is released to neutralize the acid.[38] This then initiates the chain of events described above, with results analogous to the activating effect of phenolic compounds.

C. Calcium Channels and Eicosanoids

Ca^{2+} association with release of arachidonic acid has been described by Malik[24] as a process in which α-1 adrenergic receptor activation causes increased influx of Ca^{2+} and/or mobilizes intracellular Ca^{2+}, which in turn

Phospholipase A_1

Phospholipase A_2

H_2C — fatty acid[1]

fatty acid[2] — CH

H_2C — PO_4 — BASE

Phospholipase C Phospholipase D

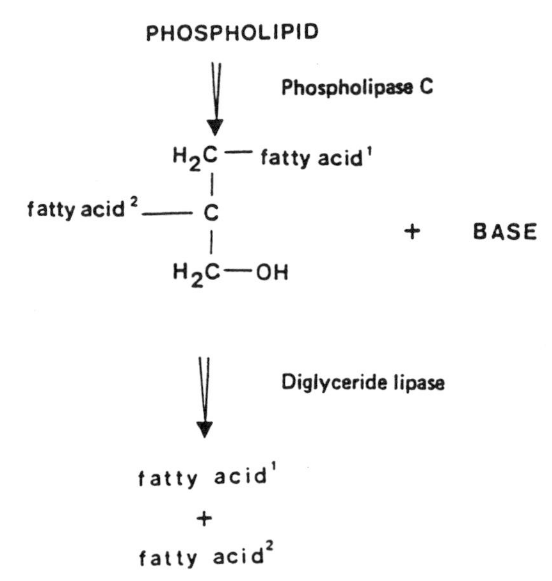

Figure 1.5 Structure of a typical phospholipid. The fatty acid attached to "hook" 1 is normally saturated, while that attached at "hook" 2 is normally unsaturated. Phospholipase A_2 cleaves specifically the unsaturated fatty acid (e.g., arachidonic acid) from the second hook. Release of arachidonic acid from phospholipids also follows the consecutive action of phospholipase C and diglyceride lipase. (From Moore, P. K., *Prostanoids: Pharmacological, Physiological and Clinical Relevance.* Cambridge University Press, New York, 1985, 7. With permission.)

binds with calmodulin and activates one or more lipases to release arachidonic acid for synthesis of eicosanoids. Furthermore, the 5-lipoxygenase enzyme is activated by Ca^{2+}.[39] Inositol lipids are associated with the Ca^{2+} signaling system via mobilization of Ca^{2+} from the endoplasmic reticulum.[40,41] Inositol 1,4,5-triphosphate has been identified as a Ca^{2+}-mobilizing second messenger, with localization of Ca^{2+} signals at discrete regions

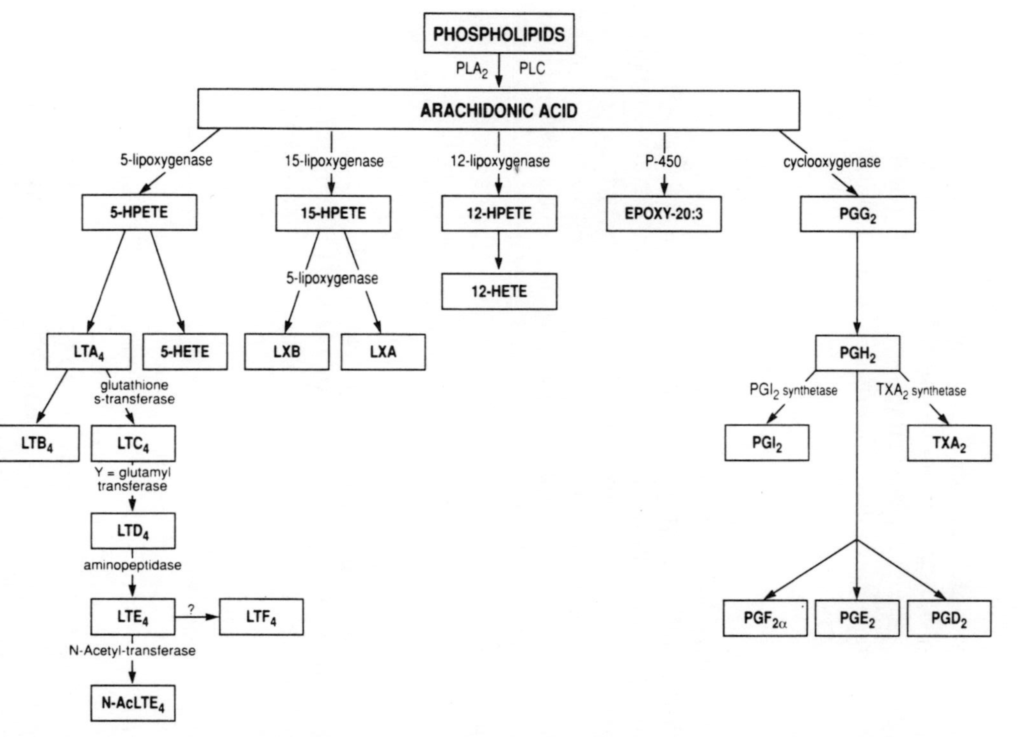

Figure 1.6 Schematic presentation of key pathways in arachidonic acid metabolism. PG = prostaglandin; LT = leukotriene; LX = lipoxin; N-$A_C$$LTE_4$ = *N*-acetyl-LTE_4; PLC = phospholipase C; PLA_2 = phospholipase A_2; HETE = hydroxyeicosatetraenoic acid; HPETE = hydroperoxyeicosatetranoic acid.

of cells and the generation of both Ca^{2+} waves and Ca^{2+} oscillations.[41] Influx of calcium ions is accordingly being recognized as an important part of the cellular events associated with inflammatory reactions.[40,42]

D. Fatty Acid Precursors

A rate-limiting factor in the synthesis of prostaglandins and leukotrienes is the quantitative dietary intake of essential fatty acids.[43-46] Arachidonic acid is the principal substrate for the cyclooxygenase and lipoxygenase pathways.[44,45] Linoleic acid, a major constituent of vegetable oils obtained from corn, safflower, soybean, and sunflower (Tables 1.1 and 1.2), may be metabolized to arachidonic acid (Figure 1.7).[46,47] In addition, γ-linolenic acid and dihomo-γ-linolenic acid can also be metabolized to arachidonic acid (Figure 1.7).[46]

Eicosapentaenoic acid ($C_{20:5}$) is a fatty acid found in fish oils which gives rise to prostaglandins, thromboxanes, and leukotrienes with five (rather than three or four) double bonds.[49] Leukotriene A_3 is derived from $C_{20:5}$.[49] A rapid biochemical and physiologic response to seafood materials is possibly due to rapid entry of eicosapentaenoate from dietary lipids into the nonesterified pools of precursor acid. The nonesterified acids can compete with arachidonate, decreasing its conversion to active prostaglandins.[48,49] However, eicosapentaenoic acid is a good substrate for lipoxygenase enzymes and can be converted to leukotrienes and other lipoxygenase products.[48] Birth weights in the Faroe Islands are among the highest in the world.[50] Danish workers[51] proposed that this is due to a high intake of marine fat, rich in eicosapentaenoic acid (n-3). They attributed the mechanism to eicosapentaenoic acid (n-3) interfering with the uterine production of dienoic prostaglandins, primarily $PGF_{2\alpha}$ and PGE_2, which are mediators of uterine contractions and cervical ripening. Fatty acid derivatives have also been reported to control ion channels, ranging from lipid-derived second messengers to signals carried via the circulation.[50,51] Two outwardly rectifying K^+ channels in the heart open in response in certain fatty acid derivatives.[52] Cyclooxygenase and lipoxygenase pathways were not found to be responsible for this action.[52,53] The suggestion has been made that fatty acids themselves activate K^+ channels through a direct action at a protein or lipid site in the membrane of smooth muscle cells, with the possibility that derivatives of the fatty acids act as second messengers.[52-54] Fatty acids found to have K^+ current activation potential include arachidonic, oleic, myristic, and linolenic acids.[53] Generation of the free fatty acids is enhanced by the action of the phospholipase enzymes described above.

E. Biologically Active Amines

Histamine has been central in discussions of allergies. Three types of histamine receptors (H_1, H_2, and H_3) are known and are associated with the *de novo* synthesis of different eicosanoids.[55-57] The receptors are found

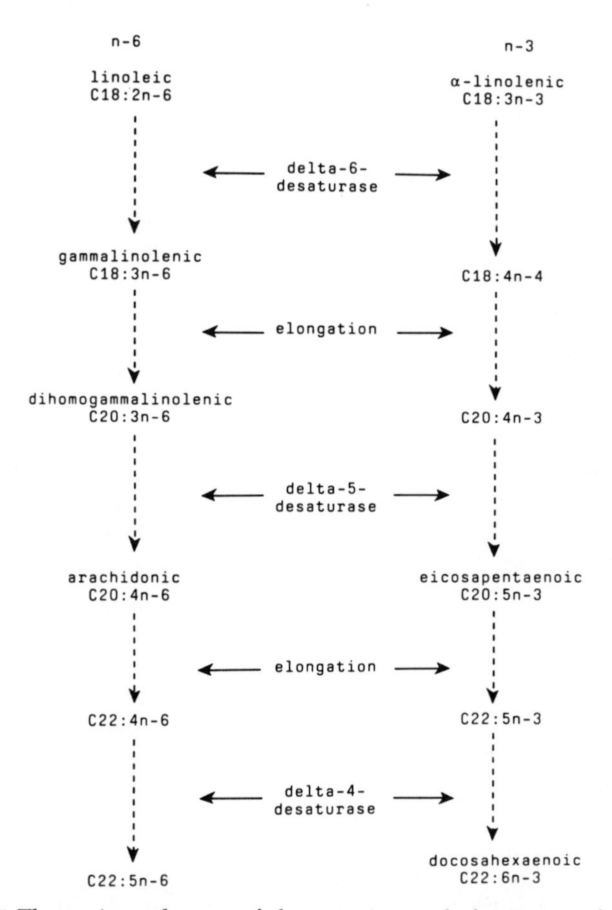

Figure 1.7 The main pathways of desaturation and elongation of the essential fatty acids. (From Fischer, S., Dietary polyunsaturated fatty acids and eicosanoid formation in humans, *Adv. Lipid Res.*, 23, 169, 1989. With permission.)

in many tissues, including the skin and gut, and also in leukocytes.[56] During the secretion of histamine by basophils and mast cells, important changes occur in the phospholipid metabolism of these cells.[55] Activation of these cells by chemicals or related compounds in sensitized individuals results in the release of arachidonic acid which parallels the release of histamine upon IgE or calcium ionophore stimulation of the cell and requires Ca^{2+} in the medium.[55-57] Ironically, blocking cell phospholipase activity also blocks histamine release, suggesting a two-pronged activation of eicosanoid production with the addition of histamine as a second stimulus via receptor stimulation as noted above.[58] Such interactions prompted Owen[58] to suggest that an alternative role for histamine might be as a comediator of inflammation.

Table 1.1
Common Sources and Effect on Plasma Lipids of Major Dietary Fatty Acids

Fatty acids		Common sources	Major effect on plasma lipids[a]
Saturated			
C12:0	Lauric	Coconut, palm kernel oil	Increases plasma total C
C14:0	Myristic	Coconut	Increases plasma total C
C16:0	Palmitic	Palm oil, beef	Increases plasma total C
C18:0	Stearic	Cocoa butter, beef	Decreases plasma total C, has no effect, raises it less than expected
Unsaturated			
Monounsaturated			
C18:1	Oleic	Olive oil, rapeseed oil, beef	Decreases plasma total C[25]
Polyunsaturated			
ω-6			
C18:2	Linoleic	Corn oil, cottonseed oil, safflower oil, soybean oil, sunflower oil	Decreases plasma total C[18,26]
ω-3			
C20:5	Eicosapentaenoic	Atlantic and king mackerel, Atlantic and Pacific herring (sardines), lake trout, chinook salmon, albacore tuna, Atlantic and sockeye salmon, bluefish, pink and chum salmon, Atlantic halibut, coho salmon, marine lipids (cod liver oil and ω-3 fatty acid supplements)	Decreases plasma triglycerides, variable LDL-C and HDL-C effects[27]
C22:6	Docosahexaenoic		

[a]LDL = low-density lipoprotein; HDL = high-density lipoprotein.

Table 1.2
Comparisons of Fatty Acid Compositions of Fats and Oils from Various Dietary Sources

	Fatty acid[a]									
Food item	<C12:0	C12:0	C14:0	C16:0	C18:0	C18:1	C18:2	C18:3	C20:5	C22:6
Fats and oils										
Beef tallow	—	0.9	3.7	24.9	18.9	36.0	3.1	0.6	—	—
Butter	7.0	2.3	8.2	21.3	9.8	20.4	1.8	1.2	—	—
Cocoa Butter	—	—	0.1	25.4	33.2	32.6	2.8	0.1	—	—
Corn oil	—	—	—	12.0	2.0	25.0	60.0	0.5	—	—
Cottonseed oil	—	—	0.8	22.7	2.3	17.0	51.5	0.2	—	—
Lard	0.1	.02	1.3	23.8	13.5	41.2	10.2	1.0	—	—
Olive oil	—	—	—	13.0	2.5	74.0	9.0	0.5	—	—
Palm kernel oil	7.2	47.0	16.4	8.1	2.8	11.4	1.6	—	—	—
Palm Oil	—	0.1	1.0	43.5	4.0	36.6	9.1	0.2	—	—
Safflower oil	—	—	—	6.5	2.5	11.5	79.0	0.5	—	—
Shortening[b]	0.2	0.4	0.4	19.3	9.9	50.6	13.5	0.6	—	—
Meat, fish, and poultry										
Beef, lean only, uncooked	—	—	0.17	1.4	0.74	2.4	0.2	0.01	—	—
Chicken, white meat, uncooked	—	—	0.01	0.3	0.1	0.4	0.2	0.01	—	—
Salmon, coho, raw	—	—	0.3	0.6	0.2	1.2	0.3	0.2	0.3	0.5
Tuna, light, canned in oil	—	—	0.03	1.4	0.1	2.8	2.7	0.07	0.03	0.1

Source: USDA Agriculture Handbook No. 8-4.[28]

[a]Values represent percent per 100 g edible portion. Only major fatty acids are presented.
[b]Soybean and palm oils, hydrogenated.

Serotonin (5-hydroxytryptamine [5-HT]) receptors (particularly the 5-HT2 receptor) stimulate the brain to release arachidonic acid via activation of phospholipase C.[59,60] The amino acid L-tryptophan in proteins is the precursor of serotonin; consequently, if serotonin receptors are overactive, some provision must be made to curb these clinical responses.

Pharmacological effects have also been attributed to other pressor amines found in plants.[61] These include such compounds as norepinephrine, synephrine, octopamine, vanillyamine, ephedrine, and tyramine.[61]

F. Casein: A Milk Protein

Casein induces the activity and release of phospholipase A_2, thereby promoting arachidonic acid metabolism by both the cyclooxygenase and lipoxygenase pathways in neutrophils and neighboring cells.[62-64] Casein also stimulates the activity of 5-lipoxygenase, the enzyme associated with the formation of leukotrienes (i.e., LTB_4, LTC_4, and LTD_4).[65]

G. Glutamic Acid and Monosodium Glutamate

The neurotransmitter glutamate activates *N*-methyl-D-aspartate receptors which stimulate phospholipase A_2 to release arachidonic acid.[66-68] The increased availability of arachidonic acid for the formation of eicosanoids is not the only physiological effect resulting from this release. Micromolar levels of arachidonic acid inhibit the uptake of glutamate into glial cells; this is not prevented by cyclooxygenase or lipoxygenase inhibitors.[67] Since glutamic acid is the precursor of γ-aminobutyric acid (GABA), a neuromodulating neurotransmitter, a deficiency of glutamic acid causes an imbalance in activating and suppressing neurotransmitters.

Glutamate is bound into the protein structure of protein-rich foods such as meat, milk, and cheese, while vegetables such as mushrooms, tomatoes, and peas have significant levels of free glutamate.[69] The free glutamate (monosodium glutamate) is added to foods because of its flavor-enhancing properties.

H. γ-Aminobutyric Acid Receptor Functions

Two different types of GABA receptors have been identified ($GABA_A$ and $GABA_B$).[70,71] The A receptor integrates Cl^- channel conductance, either inward or outward, to stabilize the cell resting potential during the activation of excitatory receptors and is therefore inhibitory.[70] B receptor activation leads to an outwardly directed K^+ current that is inhibitory, without affecting Ca^{2+} conductance.[70,71] GABA, acting through the B receptor, inhibits serotonin- and histamine H_1-stimulated inositol phospholipid turnover in the cerebral cortex.[71] Consequently, the level of GABA can have a significant effect in modulating biological responses. Zinc has been found to antagonize responses to the inhibitory transmitter effects of GABA;[72] thus, individual patient tolerance to zinc should be considered.

I. Antivitamin Effects of Some Phenolics

Antivitamin effects have been attributed to phenolic compounds.[73] Caffeic acid (Figure 1.2) has been identified as an antithiamine factor. Various other *O*-dihydric phenols cause similar inactivation of thiamine.[73] The most potent antithiamine chemicals are those with the aliphatic side chain of caffeic acid with ortho (adjacent) hydroxy groups on the benzene ring.[73] With this knowledge of antithiamine factors, a natural response is to assume that large doses of thiamine will correct the problem. Caution should be exercised, however, inasmuch as some patients may react to thiamine. In fact, some suppliers of thiamine HCl print "Warning: may cause allergic reaction" on their labels. Mills[74] found that doses of thiamine as low as 5 mg per day for 4 to 5 weeks were toxic. The side effects were said to resemble an overdose of thyroid extract and included headache, irritability, rapid pulse, and weakness.

Linatine is assumed to be an antivitamin of pyridoxal phosphate (vitamin B_6).[75] This compound was originally isolated from flax seed, but its possible presence in other plant foodstuffs is known. Caution is also advocated in using large doses of this vitamin, inasmuch as the vitamin may exert a direct toxic effect on the peripheral nervous system and might also cause seizures.[75]

Niacin is a vitamin which is a potent hypolipidemic agent, but intense flushing occurs following ingestion of pharmacologic doses of the vitamin.[76] Knowing that niacin-induced flushing can be substantially attenuated by pretreatment with cyclooxygenase inhibitors, Morrow et al.[76] assumed that the vasodilatation was mediated by a prostaglandin, which prompted them to investigate that hypothesis. Intense flushing occurred in three normal volunteers following ingestion of 500 mg of niacin. They also noted extreme fatigue and sleepiness, and two experienced a feeling of dissociation from their surroundings. Plasma levels of 9α- and 11β-PGF_2 increased dramatically by 800-, 430-, and 535-fold, respectively, in the three volunteers. Since these prostaglandins are metabolites of PGD_2, they concluded that PGD_2 is the prostaglandin for inducing vasodilatation in humans. Release of PGD_2 is not accompanied by a release of histamine, suggesting that the origin of the prostaglandin is not the mast cell. In a secondary study, they pretreated one of the volunteers with 200 mg of indomethacin (a cyclooxygenase inhibitor) daily for 4 days. This treatment markedly reduced flushing and reduced plasma prostaglandin levels by 95%.

J. Animal Studies Using Phenolics

The pharmacological effects of phenolic compounds have also been demonstrated in animals.[77] In a pilot study, four calves were fed whole cow's milk to which benzyl (phenyl) isothiocyanate was added to duplicate the daily quantity (90 mg) consumed by calves fed a soybean milk replacer. These calves exhibited the same pattern of diarrhea and illness as calves fed a soybean milk replacer. Total gains from 3 days to 4 weeks of age averaged 0.7 kg, whereas calves concurrently being fed regular cow's milk gained 8.0 kg and had normal bowel function. Other physiological responses were measured in a subsequent study with 60 calves.[77] Three treatments were used; these included cow's milk as a control, soybean milk replacer extracted with ethanol, and soybean milk replacer extracted with hexane. Ethanol extraction was more effective than hexane in removing phenolics (2.2 vs. 1.0% remaining in replacer). Tachycardia and bronchoconstriction were responses observed which are typical of the pharmacological effects of phenolic compounds and eicosanoids. For example, heart rates (beats per minute) averaged 88 (milk), 99 (ethanol extracted), and 116 (hexane extracted). Mortality rates were 0/20, 4/20, and 9/20 for the respective treatments. Diarrhea was profuse in the calves fed the soybean milk replacers. Degenerative changes and inflammation in the midsections of the small intestine were noted upon postmortem examination. Serum $PGF_{2\alpha}$ was elevated in calves fed the soybean milk replacers vs. cow's milk.

K. Kinins in Formation of Eicosanoids

Kallikrein is an enzyme occurring in urine, plasma, and some glands which acts on a plasma protein precursor to form bradykinen (Figure 1.1).[7] Like angiotensin II, bradykinin stimulates eicosanoid biosynthesis in several perfused organs (including the kidney) by activating membrane phospholipase A_2 activity (Figure 1.5).[33]

Many of the pharmacological effects of bradykinin in the body may be altered in character or abolished completely by nonsteroidal anti-inflammatory drugs (NSAID), indicating that they are mediated by eicosanoids.[33] In the dog kidney, indomethacin significantly reduced the dilatation induced by bradykinin.[33] Bradykinin has a selective effect on sodium and chloride transport. Tomita et al.[78] propose that the electroneutral transport process affected by bradykinin most likely involves coupled transport of sodium and chloride.

L. Effects of Inflammatory Agents on Organ Systems

Clinical effects of eicosanoid chemicals on different organ systems are discussed in chapters which follow. Therapeutic use of phenolic compounds to prevent, or arrest, these inflammatory responses will be discussed in Chapter 14.

REFERENCES

1. **Aas, K.,** The biochemistry of food allergens: what is essential for future research?, In *Food Allergy,* Reinhardt, D. and Schmidt, E., Eds., Raven Press, New York, 1988, 1.
2. **Hubbard, R.,** Fundamentals of immunology and immunoglobulins, in *Immunotoxicology,* Gibson, G. G., Hubbard, R., and Parke, D. V., Eds., Academic Press, New York, 1983, 5.
3. **Roberts, R. L.,** The hyperimmunoglobulin E syndrome: clinical presentation and immunologic dysfunction, *Insights Allergy,* 4, 1, 1989.
4. **Soothill, J. F.,** Genetic and nutritional variations in antigen handling and disease, in *Immunology of the Gut,* Elsevier, Amsterdam, 1977, 225.
5. **Edwards, R. G.,** Structural determinants of drug allergenicity: implication in drug design, in *Immunotoxicology,* Gibson, G. G., Hubbard, R., and Parke, D. V., Eds., Academic Press, New York, 1983, 385.
6. **Ring, J.,** Dermatologic diseases secondary to food allergy and pseudoallergy, in *Food Allergy,* Reinhardt, D. and Schmidt, E., Eds., Raven Press, New York, 1988, 271.
7. **Chandra, R. K. and Shah, A.,** Immunologic mechanisms, in *Food Intolerance,* Chandra, R. K., Ed., Elsevier, New York, 1984, 55.
8. **Piper, P. J., Sampson, A. P., Yaacob, H. B., and McLeod, J. M.,** Leukotrienes in the cardiovascular system, in *Advances in Prostaglandin, Thromboxane, and Leukotriene Research,* Vol. 20, Samuelsson, B., Dahlen, S. -E., Fritsch, J., and Hedqvist, P., Eds., Raven Press, New York, 1990, 146.
9. **Gardner, R. W., McGovern, J. J., and Brenneman, L. D.,** The role of plant and animal phenyls in food allergy, paper presented at the 37th Annu. Congr. American College of Allergists, Washington, D.C., April 4-8, *(Note: results are included in the appendix of this book.)*
10. **Burn, J. H.,** Some important phenolic compounds in pharmacy, in *The Pharmacology of Plant Phenolics,* Fairbairn, J. W., Ed., Academic Press, New York, 1959, 1.
11. **Singleton, V. L.,** Naturally occurring food toxicants: phenolic substances of plant origin common in foods, *Adv. Food Res.,* 27, 149, 1981.
12. **Archer, D. L.,** Immuntoxicology of foodborne substances: an overview, *J. Food Prot.,* 41, 983, 1978.
13. **Ames, B. N., Profet, M., and Gold, L. S.,** Dietary pesticides (99.99% all natural), *Proc. Natl. Acad. Sci. U.S.A.,* 87, 7777, 1990.
14. **Ames, B. N., Profet, M., and Gold, L. S.,** Nature's chemicals and synthetic chemicals: comparative toxicology, *Proc. Natl. Acad. Sci. U.S.A.,* 87, 7782, 1990.
15. **Lockey, S. D., Sr.,** Hypersensitivity to tartrazine (FD&C yellow no. 5) and other dyes and additives present in food and pharmaceutical products, *Ann. Allerg.,* 46, 81, 1981.
16. **Desmond, R. E. and Trautlein, J. J.,** Tartrazine (FD&C yellow #5) anaphylaxis: a case report, *Ann. Allerg.,* 38, 81, 1981.
17. **Loblay, R. H. and Swain, A. R.,** Adverse reactions to tartrazine, *Food Technol. Aust.,* 37, 508, 1985.
18. **Coulombe, R. A., Jr. and Sharma, R. P.,** Neurobiochemical alterations induced by the artificial sweetener aspartame (NutraSweet), *Toxicol. Appl. Pharmacol.,* 83, 79, 1986.

19. **Johns, D. R.,** Migraine provoked by aspartame, *N. Engl. J. Med.,* 315, 456, 1986.
20. **Cody, V., Middleton, E., Jr., Harborne, J. B., and Beretz, A.,** Eds., *Plant Flavonoids in Biology and Medicine: Biochemical, Pharmacological and Structure-Activity Relationships,* Alan R. Liss, New York, 1986.
21. **Cody, V., Middleton, E., Jr., Harborne, J. B., and Beretz, A.,** Eds., *Plant Flavonoids in Biology and Medicine II: Biochemical, Cellular, and Medicinal Properties,* Alan R. Liss, New York, 1988.
22. **Lands, W. E. M. and Hanel, A. M.,** Inhibitors and activators of prostaglandin biosynthesis, in *Prostaglandins and Related Substances,* Pace-Asciak, C. and Granstom, E., Eds., Elsevier, New York, 1983, chap. 6.
23. **DeEds, F.,** Physiological effects and metabolic fate of flavonoids, in *The Pharmacology of Plant Phenolics,* Fairbairn, J. W., Ed., Academic Press, New York, 1959, 91.
24. **Malik, K. V.,** Interaction of arachidonic acid metabolites and adrenergic nervous system, *Am. J. Med. Sci.,* 295, 280, 1988.
25. **Bhattacharya, S. K., Dasgupta, G., and Sen, A. P.,** Prostaglandins modulate central serotonergic neurotransmission, *Indian J. Exp. Biol.,* 27, 393, 1989.
26. **Ganong, W. F.,** *Review of Medical Physiology,* 9th ed., Appleton & Lange, East Norwalk, CT, 1979, chap. 1.
27. **Lewis, R. A.,** Mast cell-dependent immediate hypersensitivity responses, in *Plant Flavonoids in Biology and Medicine: Biochemical, Pharmacological, and Structure-Activity Relationships,* Cody, V., Middleton, E., Jr., and Harborne, J. R., Eds., Alan R. Liss, New York, 1986, 457.
28. **Bonventre, J. V.,** Calcium in renal cells. Modulation of calcium-dependent activation of phospholipase A_2, *Environ. Health Perspect.,* 84, 155, 1990.
29. **Cowell, A. M. and Buckingham, J. C.,** Eicosanoids and the hypothalamo-pituitary axis, *Prostaglandins Leukotrienes Essential Fatty Acids Rev.,* 36, 235, 1989.
30. **Janniger, C. K. and Ralcis, S. P.,** The arachidonic acid cascade: an immunologically based review, *J. Med.,* 18, 69, 1987.
31. **Axelrod, J., Burch, R. M., and Jelsema, C. L.,** Receptor-mediated activation of phospholipase A_2 via GTP-binding proteins: arachidonic acid and its metabolites as second messengers, *Trends Neurosci.,* 11, 117, 1988.
32. **Vergroesen, A. J.,** Hypothesis for interactions between acetylcholine and prostaglandin biosynthesis: an introduction, in *Nutrition and the Brain,* Vol. 5, Barbeau, A., Growdon, J. H., and Wurtman, R. J., Eds., Raven Press, New York, 1979, 109.
33. **Moore, P. K.,** *Prostanoids: Pharmacological, Physiological and Clinical Relevance,* Cambridge University Press, New York, 1985.
34. **Dhillon, M., Roberts, C., Nunn, T., and Kuo, M.,** Mapping human T cell epitopes on phospholipase A_2: the major bee venom, *J. Allergy Clin. Immunol.,* 90, 42, 1992.
35. **Schumacher, M. J., Schmidt, J. O., Egen, N. B., and Dillon, K. A.,** Biochemical variability of venoms from individual European and Africanized honeybees *(Apis mellifera), J. Allergy Clin. Immunol.,* 90, 59, 1992.
36. **Blecher, M., Merlin, N. S., and Ro'ane, J. T.,** Control of the metabolism and lipolytic effects of cyclic 3′-5′-adenosine monophosphate in adipose tissue by insulin, methyl xanthines and nicotinic acid, *J. Biol. Chem.,* 243, 3973, 1968.

37. **Lewin, S.,** *Vitamin C: Its Molecular Biology and Medical Potential,* Academic Press, New York, 1976.
38. **Meszaros, L. S.,** The biochemistry of stress, *Chemtech,* 18, 223, 1988.
39. **Jakschik, B., Sun, F. F., Lee, L. -H., and Steinhoff, M. M.,** Calcium stimulation of a novel lipoxygenase, *Biochem. Biophys. Res. Commun.,* 95, 103, 1980.
40. **Farese, R. V.,** Calcium as an intracellular mediator of hormone action: intracellular phospholipid signaling systems, *Am. J. Med. Sci.,* 296, 223, 1988.
41. **Berridge, M. J.,** Inositol triphosphate, calcium, lithium, and cell signaling, *J.A.M.A.,* 262, 1834, 1989.
42. **Ahmed, T., Kim, C. S., and Danta, I.,** Inhibition of antigen-induced bronchoconstriction by a new calcium antagonist, gallopamil: comparison with cromolyn sodium, *J. Allergy Clin. Immunol.,* 81, 852, 1988.
43. **Higgs, G. A.,** The effects of dietary intake of essential fatty acids on prostaglandin and leukotriene synthesis, *Proc. Nutr. Soc.,* 44, 181, 1985.
44. **Land, W. E. M.,** The biosynthesis and metabolism of prostaglandins, *Annu. Rev. Physiol.,* 41, 633, 1979.
45. **Oliw, E., Granström, E., and Ånggård, E.,** The prostaglandins and essential fatty acids, in *Prostaglandins and Related Substances,* Pace-Asciak, C. and Granström, E., Eds., Elsevier, New York, 1983, 1.
46. **Fischer, S.,** Dietary polyunsaturated fatty acids and eicosanoid formation in humans, *Adv. Lipid Res.,* 23, 169, 1989.
47. **Carter, J. P.,** Gamma-linolenic acid as a nutrient, *Food Technol.,* 42, 72, 1988.
48. **Lands, W. E. M. and Kulmacz, R. J.,** The regulation of the biosynthesis of prostaglandins and leukotrienes, *Prog. Lipid Res.,* 25, 105, 1986.
49. **Palmer, R. M. J. and Salmon, J. S.,** Inhibition of 5′ lipoxygenase: relevance to inflammation, in *Drugs Affecting Leukotrienes and Other Eicosanoid Pathways,* Ser. A, Vol. 95, Samuelsson, B., Berti, F., Folco, G. C., and Velvo, G. P., Eds., Plenum Press, New York, 1985, 311.
50. **Olsen, S. F. and Joensen, H. D.,** High liveborn birthweights in the Faroes: a comparison between birthweights in the Faroes and in Denmark, *J. Epidemiol. Commun. Health,* 39, 27, 1985.
51. **Olsen, S. F., Hansen, H. S., Sorensen, T. I. A., Jensen, B., Secher, N. J., Somner, S., and Knudsen, L. B.,** Intake of marine oil, rich in (n-3)-polyunsaturated fatty acids, may increase birthweight by prolonging gestation, *Lancet,* 2, 367, 1986.
52. **Kim, D. and Clapham, D. E.,** Potassium channels in cardiac cells activated by arachidonic acid and phospholipids, *Science,* 244, 1174, 1989.
53. **Ordway, R. W., Walsh, J. V., Jr., and Singer, J. J.,** Arachidonic acid and other fatty acids directly activate potassium channels in smooth muscle cells, *Science,* 244, 1176, 1989.
54. **Parke, D. V. and Gibson, G. C.,** Molecular mechanism of chemical-mediated immunopathology, in *Immunotoxicology,* Gibson, G. G., Hubbard, R., and Parke, D. V., Eds., Academic Press, New York, 1983, 5.
55. **Douglas, J. S. and Brink, C.,** 2. Mediators — histamine and prostanoids, *Am. Rev. Respir. Dis.,* 136, S21, 1987.
56. **Hill, S. J.,** Distribution, properties, and functional characteristics of three classes of histamine receptors, *Pharmacol. Rev.,* 42, 46, 1990.
57. **Ninnemann, J. L.,** *Prostaglandins, Leukotrienes and the Immune Response,* Cambridge University Press, New York, 1988, 194.

58. **Owen, D. A. A.,** Inflammation — histamine and 5-hydroxytryptamine, *Br. Med. Bull.,* 43, 256, 1987.
59. **Conn, P. J. and Sanders-Bush, E.,** Central serotonin receptors: effector systems, physiological roles and regulation, *Psychopharmacology,* 92, 267, 1987.
60. **Sanders-Bush, E., Tsutsumi, M., and Burris, K. D.,** Serotonin receptors and phosphatidylinositol turnover, in *The Neuropharmacology of Serotonin,* Whitaker-Azmitia, P. and Peroutka, S. J., Eds., *Ann. N.Y. Acad. Sci.,* Vol. 600, 1990.
61. **Robinson, T.,** *The Organic Constituents of Higher Plants,* 4th ed., Cordus Press, North Amherst, MA, 1980.
62. **Chang, H. W., Kudo, I., Hara, S., Karawawa, K., and Inoue, K.,** Extracellular phospholipase A_2 activity in peritoneal cavity of casein treated rats, *J. Biochem.,* 100, 1099, 1986.
63. **Chang, H. W., Kudo, I., Tomita, M., and Inoue, K.,** Purification and characterization of extracellular phospholipase A_2 from peritoneal cavity of caseinate treated rats, *J. Biochem.,* 102, 147, 1987.
64. **Miller, M. J. S., Withely, S. A., and Clark, D. A.,** Casein: a milk protein with diverse biological consequences (minireview), *Proc. Soc. Exp. Biol. Med.,* 195, 143, 1990.
65. **Chang, W. C. and Su, G. W.,** Stimulation of 5-lipoxygenase activity in polymorphonuclear leukocytes by casein treatment, *Biochem. Pharmacol.,* 36, 3303, 1987.
66. **Barbour, B., Szatkowski, M., Ingledew, N., and Attwell, D.,** Arachidonic acid induces a prolonged inhibition of glutamate uptake into glial cells, *Nature,* 342, 918, 1989.
67. **Duruis, A., Sebben, M., and Haynes, L.,** NMDA receptors activate the arachidonic acid cascade system in striatal neurons, *Nature,* 336, 68, 1988.
68. **Lazarewicz, J. W., Wroblewski, J. T., Palmer, M. E., and Costa, E.,** Activation of *N*-methyl-D-aspartate-sensitive glutamate receptors stimulates arachidonic acid release in primary cultures of cerebellar granule cells, *Neuropharmacology,* 27, 765, 1988.
69. **Food Technologists' Expert Panel on Food Safety and Nutrition,** Monosodium glutamate (MSG), *Food Technol.,* 41, 143, 1987.
70. **Bormann, J.,** Electrophysiology of $GABA_A$ and $GABA_B$ receptor subtypes, *Trends Neurosci.,* 11, 112, 1988.
71. **Ogata, N.,** Pharmacology and physiology of $GABA_B$ receptors (minireview), *Gen. Pharmacol.,* 21, 395, 1990.
72. **Westbrook, G. L. and Mayer, M. L.,** Micromolar concentrations of Zn^{2+} antagonize NMDA (*N*-methyl-D-aspartate) and GABA responses of hippocampal neurons, *Nature,* 328, 640, 1987.
73. **Somogyi, J. C.,** Antivitamins, in *Toxicants Occurring Naturally in Foods,* 2nd ed., National Academy of Sciences, Washington, D.C., 1973, 254.
74. **Mills, C. A.,** Thiamine overdose and toxicity, *J.A.M.A.,* 116, 2101, 1941.
75. **Alhadeff, L., Gualtieri, C. T., and Lipton, M.,** Toxic effects of water-soluble vitamins, *Nutr. Rev.,* 42, 33, 1984.
76. **Morrow, J. D., Parson, W. G., III, and Roberts, L. J., II,** Release of markedly increased quantities of prostaglandin D_2 *in vivo* in humans following the administration of nicotinic acid, *Prostaglandins,* 38, 263, 1989.
77. **Gardner, R. W., Shupe, M. G., Brimhall, W., and Weber, D. J.,** Causes of adverse responses to soybean milk replacers in young calves, *J. Dairy Sci.,* 73, 1312, 1990.

78. **Tomita, K., Ujüe, K., Maeda, Y., Iino, Y., Yoshiyama, N., and Shiigai, T.,** Effects of mineralocorticoid on kinase activity along the distal nephron segments of the rat kinins V. Part A, Abe, K., Moriya, H., Fujii, S., Eds., *Adv. in Exp. Med. Biol.,* 247A, 181, 1989.

2

Gastrointestinal Disturbances

I. INTRODUCTION

One third of the population suffers from adverse reactions to foods, according to a survey of a university teaching staff in Britain.[1] Plasma IgE was measured in a random subset of 99 women with food intolerances and was found to be significantly lower in those with major problems than in the rest.[2] This suggested to Burr and Merrett[2] that allergy is probably not a common cause of food intolerance, a viewpoint shared by Anderson.[3] Sampson[4] recognized that an "adverse reaction" to a food may be categorized into either "food hypersensitivity" (i.e., immunological allergic reaction) or "food intolerance" reactions. He also proposed that food intolerance probably accounts for most adverse food reactions. Sampson[4] subcategorized causes of food intolerance as "food toxicity (poisoning)", "idiosyncracy", "pharmacologic reaction", or "metabolic reaction".

Vomiting, diarrhea, nausea, colic, abdominal pain, colitis, abdominal gas, constipation, flatulence,and a feeling of fullness are all symptoms of gastrointestinal reactions to foods and food additives.[3,5-8]

II. EICOSANOIDS IN GASTROENTERITIS

Administration of prostaglandins E_2 or $F_{2\alpha}$ (Figure 2.1) produces the gastrointestinal symptoms cited above.[5,9-11] These were the first side effects noted when prostaglandins were employed in gynecological practice.[11] Endogenous prostaglandins may be the activating factors associated with gastrointestinal disorders following the ingestion of specific foodstuffs.[5] Buissert et al.[5] were able to prevent symptoms of food intolerance in five out of six patients by using prostaglandin-synthesis inhibitors. High concentrations of prostaglandins were found in the stools of the food-sensitive patients if inhibitors were not used. Total IgE concentrations were not raised. Their findings are in accord with the view that prostaglandins are involved in the mediation of abnormal increases in intestinal motility and secretions.[8-10] Jones et al.[12] referred to this abnormal bowel action as irritable bowel syndrome, in which specific foods were found to provoke symptoms in 14 of 21 patients. No difference was detected in plasma

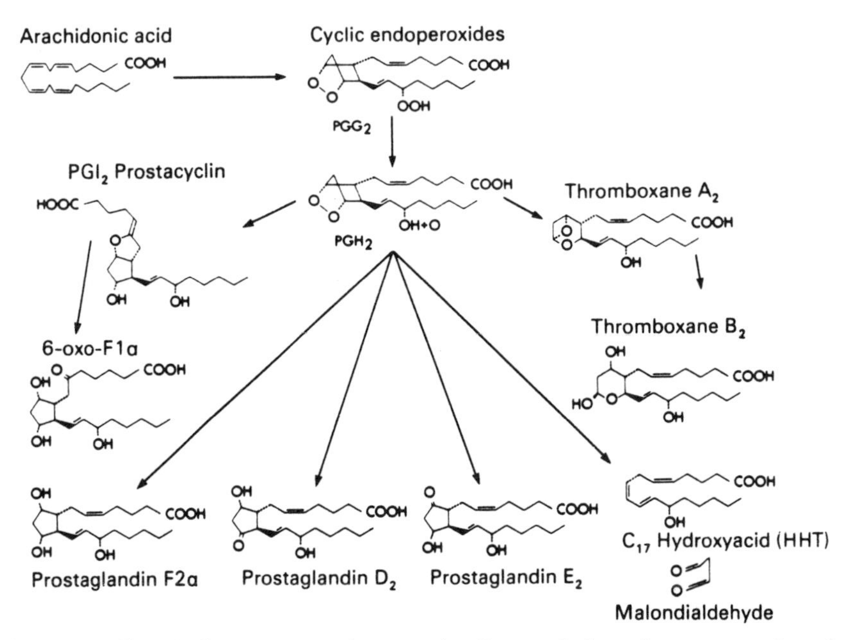

Figure 2.1 Chemical structures of prostaglandins and thromboxanes produced via the cyclooxygenase pathway.

glucose, histamine, or immune complexes produced after challenge or control foods. Rectal prostaglandin E_2 (PGE_2), however, increased significantly, and, in an additional five patients, rectal PGE_2 correlated with wet fecal weight. They proposed that food intolerance associated with prostaglandin production is an important factor in the pathogenesis of irritable bowel syndrome.

Increased generation of leukotrienes (particularly LTB_4) has been detected in the inflamed mucosa of patients with ulcerative colitis.[13] A drug selective for inhibition of the 5-lipoxygenase pathway has proven effective in the treatment of ulcerative colitis.[13]

III. GASTROINTESTINAL MOTILITY

Prostaglandins have a contractile effect on the smooth muscles of the gastrointestinal tract.[8,10,14] They mediate their physiological effects through interactions with specific receptor sites on target cells.[15] Some receptors for prostaglandins are coupled with adenylate cyclase, thus stimulating synthesis and accumulation of cyclic adenosine monophosphate (cAMP) in cells.[16] The cAMP acts as a second messenger within synapses. This results in a closure of a crucial potassium channel in the presynaptic membrane and allows more calcium to flow into the synapse through voltage-dependent calcium channels, leading to contraction of blood vessels as well as smooth muscles of the intestines and bronchioles.[16] Evidence of the role of calcium ions in irritable bowel syndrome was offered by Narducci et al.[17] Octylonium bromide, a smooth muscle relaxant which acts by interfering with calcium ion mobilization, was given as a treatment to ten patients with irritable bowel syndrome.[17] Treatment reduced their colonic response to eating to a very short increase in colonic activity limited to the first 30 minutes.

IV. ELECTROLYTE AND WATER MOVEMENT

Besides effects on gastrointestinal motility, the eicosanoids also influence water and electrolyte transport across the intestinal mucosa, thus causing a more watery stool.[11,18] Crypt cells in the intestine are primarily responsible for secretion of fluid, while villus cells absorb fluids. Attention is directed to the finding that in these two cell types intracellular cAMP and Ca^{2+} play key roles in controlling the rate of secretion and absorption, respectively, of Na^+ and Cl^-.[11,18] The eicosanoids produced in response to cAMP and Ca^{2+} have been shown to stimulate intestinal Cl^- and fluid secretion and to inhibit intestinal NaCl and fluid absorption.[18] A net increase in secretions into the lumen follows.[11,18] Human cholera is associated with an increased luminal "overflow" of PGE_2, with a substantial increase in fluid secretion.[17] Fluid secretion can be decreased by indomethacin or by the serotonin (5-HT) antagonist receptor, ketanserin, suggesting that cholera toxin stimulates the release of 5-HT, which in turn causes the release of PGE_2.[18]

V. PHENOLIC COMPOUNDS IN GASTROENTERITIS

Inasmuch as prostaglandins have been identified in gastrointestinal disturbances, one must then identify factors which increase the production of those compounds. Tseng et al.[19] tested patients with a group of phenolic compounds (i.e., caffeic acid, ferulic acid, coumaric acid, homovanillic acid, protocatechuic acid, and related phenolics). They found that the phenolic compounds stimulate the biosynthesis of prostaglandins. These phenolic compounds are found in plant materials, including foodstuffs and pollens.[20-23]

Some plant phenolics, on the other hand, *inhibit* prostaglandin biosynthesis. They include such chemical groups as *N-trans-* and *N-cis-*feruloyltyramines, along with scopoletin, umbelliferone, *N*-cinnamoyldopamine, *N*-caffeoyl-*e*-phenethylamine, and coniferyl alcohol.[22] Such an effect is in the same category as aspirin as a cyclooxygenase inhibitor. This in turn could result in a detouring of arachidonic acid to synthesis of leukotrienes via the lipoxygenase pathway (Figure 2.2). This concept is supported by an opposite effect, i.e., caffeic acid is known to inhibit lipoxygenase activity, yet reportedly stimulates prostaglandin biosynthesis.[19,24]

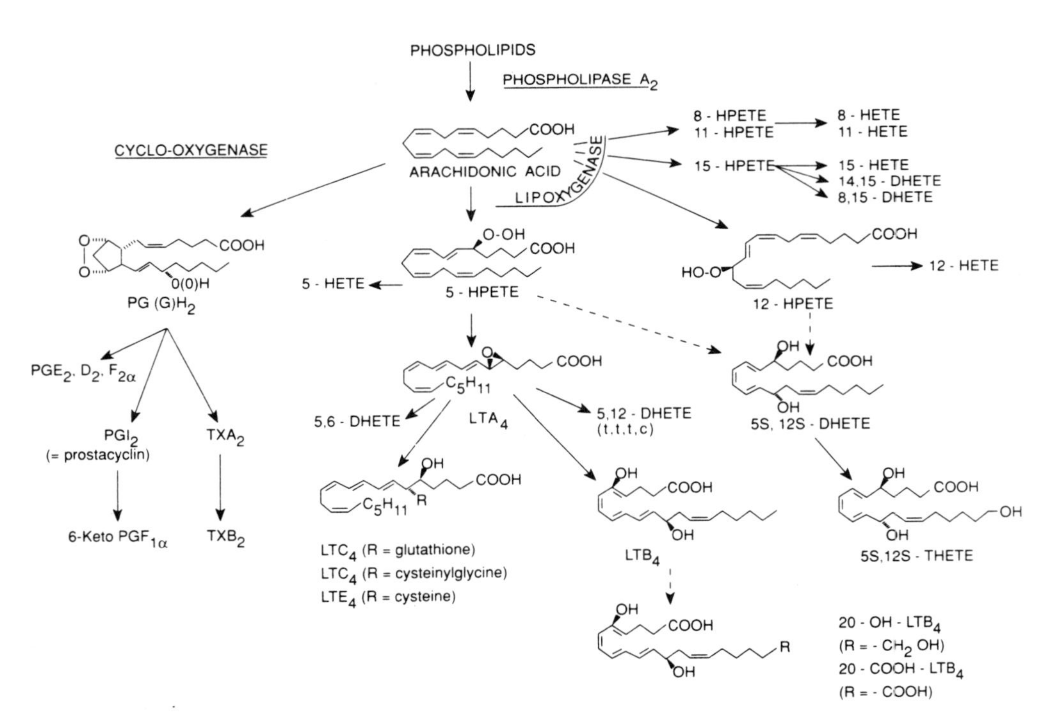

Figure 2.2 Alternative pathways in the conversion of arachidonic acid to eicosanoids. Blockage of one pathway may lead to increased synthesis via the second pathway.

Verification of the role of phenolics in increasing production of prostaglandins also came from a study of 100 patients with serious adverse responses to foods and chemical inhalants[25] (see also appendix). Laboratory analysis revealed elevated $PGF_{2\alpha}$, decreased IgE (RAST usually negative), decreased IgA, elevated immune complexes, and decreased complement. These patients were then tested with 12 phenolic compounds identified in foods by sublingual application in a placebo-controlled test. Systemic reactions duplicated the reaction symptoms resulting from ingestion of foodstuffs.

Immunotoxicology of phenolic compounds in foodborne substances has also been proposed as a possible reason for gastrointestinal allergies, toxigenic diarrhea, and pathogenic invasion through the gut wall.[28] Such phenolics as gallic acid (tannins), vanillin, methylparaben, *p*-hydroxybenzoic acid, and vanillic acid were investigated, as well as food additives such as butylated hydroxyanethole (BHA) and butylated hydroxytoluene (BHT), and were found to be immunosuppressive. The Mishell-Dutton system was used to quantitate antibody-forming cells.[26] Gallic acid blocked functions of T-lymphocytes that were dependent on the presence of functional macrophages; both helper T-cell function and nitrogen-induced suppressor T-cell function were shown to be macrophage dependent and therefore susceptible to gallic acid-induced suppression.[26] Another effect of toxic chemicals in the gastrointestinal tract is cAMP reversal of induction of suppressor T-lymphocytes (a normal immunoregulatory process).[26] Accordingly, a further detrimental effect of elevated prostaglandins could result because of activating synthesis of more cAMP (see Chapter 1 for explanation).

Blalock et al.[22] reported that foodborne phenolic compounds inhibited antibody production and also suppressed clone formation. They likened the immunosuppressive action of the phenolic agents with that of interferon. They suggested that these immunosuppressive and anticellular activities could not be due to toxicity since the inhibitory concentrations were far lower than those which are toxic (i.e., microgram per milliliter quantities). A different role of phenolic substances of plant origin in animals has been proposed.[27,28] Toxicity and related physiological activity of these compounds were emphasized by these investigators. They cited toxicity due to synergism, bonding with body polymers, interference with metabolism of normal phenols (catecholamines, tyrosine, vitamin K), and involvement of the skin and liver uncoupling of oxidative phosphorylation, production of glucosuria (phloridzin), and photosensitization (psoralen).[27]

VI. ABDOMINAL GAS

Patient complaints of distress, or discomfort, from excessive abdominal gas are commonly associated with food intolerances. The abnormality is likely due to a motility disorder.[6] Accelerated motility of the intestine would be expected to provide colonic bacteria with a larger supply of undigested carbohydrate for subsequent fermentation and related gas production. Identifying the causative food (or related chemicals) and either treatment for, or avoidance of, the active agents offers the best alternative for relief.

Some patients also experience a chronic belch after eating certain foods.[6] The patient frequently identifies flavors or odor components in the eructated gas which are peculiar to the food eaten.[25] These are the aromatic compounds which give flavor to the food and which may be the agents activating the response.

Flatulence effects of soy flour are reduced when it is extracted with 80% ethanol.[29] A chemical factor in soy which is readily extracted by ethanol is the isoflavone genistein (biochanin A), which also has estrogenic activity.[29] This is postulated as one of the factors in soy associated with flatulence,[25,29] along with benzyl isothiocyanate, benzyl thiocyanate, and a number of other phenolic compounds in soybeans which are soluble in alcohol.

VII. AMINES

Physiologically active amines in common fruits and vegetables as well as in fermentation products such as cheeses, yeast extract, alcoholic drinks, and meat products have been noted for their clinical effects.[3,30,31] Histamine, serotonin, tyramine, tryptamine, dopamine, and norepinephrine have all been identified in certain foodstuffs.[30-33] These compounds may exert a pharmacological effect if consumed in sufficient quantities, including being a stimulus for production of abnormal amounts of metabolites of arachidonic acid (prostaglandins, leukotrienes, and thromboxanes). The initial finding of serotonin, dopamine, and norepinephrine in bananas[33] prompted research into other foodstuffs. Documented pharmacological effects of serotonin (5-HT) include inhibition of gastric acid secretion, stimulation of mucus output, stimulation of intestinal secretion, and both excitatory and inhibitory effects on motor function of the bowel.[34] 5-HT also has a primary role in the emetic reflex and associated nausea.[35] Symptoms of nausea and vomiting are alleviated by the administration of the serotonin synthesis inhibitor *p*-chlorophenylalanine.[36] The 5-HT_3 receptor has been identified as the receptor associated with the emetic reflex inasmuch as antagonists of this receptor can abolish vomiting.[35]

VIII. ACETYLCHOLINE

Acetylcholine is a neural transmitter which may also increase intestinal motility.[37] Lecithin (phosphatidylcholine) from soybeans and egg yolk enhances the production of acetylcholine by supplying an abundance of choline. Solanidine found in potatoes is an inhibitor of acetylcholine esterase. The physiological effects of this toxin includes diarrhea due to an elevation in serum acetylcholine.[38] Acetylcholine causes Ca^{2+} release from an intracellular Ca^{2+} store in intestinal smooth muscle.[38] This can cause activation of phospholipase A_2, release of fatty acids, and hence production of eicosanoids which activate contraction of smooth muscles of the gut.[39,40]

IX. XANTHINES

Potatoes, yeast, animal organs, coffee beans, and tea are sources of xanthines,[41] which have a pharmacological effect on the gastrointestinal tract resulting in abnormal motility patterns.[42]

X. FOOD ADDITIVES

Potassium bromate is an additive used as a bread and flour improving agent.[43] Ingestion of the bromate can cause vomiting and diarrhea.[41]

Ethylene gas occurs naturally in ripening fruit and is also applied artificially by the food industry to hasten the ripening process.[21,42] Ethylene is metabolized by plants to ethylene glycol, ethylene oxide, and carbon dioxide.[21] Ethylene glycol causes vomiting, respiratory failure, and other toxic effects.[42] Ethylene oxide is also used as a fumigant for foodstuffs. The oxide is likewise highly irritating to mucous membrane.[42]

XI. STEROIDAL HORMONE EFFECTS

A peculiar "functional" bowel disease exacerbated during the post-ovulatory phase of the menstrual cycle in women is obviously related to hormonal causation.[44] Abdominal pain, daily nausea, intermittent vomiting, and altered stool habits characterize this particular disorder.[44] Mathias et al.[44] administered a gonadotropin-releasing hormone agonist, leuprolide acetate (Lupron®), once daily (0.5 mg subcutaneously) for 15 months to affected patients. This dose controlled pain, nausea, and vomiting. Withdrawing leuprolide induced recurrence of symptoms within 3 to 5 days. These investigators speculated that increased levels of luteinizing hormone (due to injection of leuprolide acetate) might antagonize intestinal smooth muscle or enteric neurons, or the leuprolide acetate might have a direct effect on down-regulation of the enteric neurons and/or smooth muscle.

Progesterone, estrogen, testosterone, and other steroids associated with reproduction could be involved in such gastrointestinal disorders, as discussed above. Sitosterol is yet another sterol which should be considered as causative because of its presence in many plant foodstuffs.[21,45]

XII. MICROBIAL TOXINS ASSOCIATED WITH DIARRHEA

Diarrhea from *Vibrio cholerae* results from secretion of the enterotoxin designated as choleragen. This toxin stimulates the enzyme adenyl cyclase, which converts ATP to cAMP.[46] cAMP in turn stimulates secretion of Cl^- and inhibits absorption of Na^+, resulting in a copious fluid loss and an electrolyte imbalance. An identical response is initiated by a different toxin from *Escherichia coli*.[46] Increased contraction of the smooth muscles of the intestine in association with increased cAMP and Ca^{2+} influx of course adds to this disorder, as discussed earlier.

Excessive intake of ascorbic acid is known to cause diarrhea via the same biochemical mechanism as the response to microbial toxins.[47] The explanation is ascorbic acid enhances higher levels of availability of norepinephrine by direct production, and it does so indirectly by protection of the norepinephrine from degradation.[48] Norepinephrine (or epinephrine) activates adenyl cyclase, which in turn potentiates formation of cAMP as well as inactivating phosphodiesterase, which catalyzes the hydrolysis of cAMP.[48] This results in elevated levels of cAMP with concomitant increases in production of eicosanoids.

REFERENCES

1. **Bender, A. E. and Matthews, D. R.,** Adverse reaction to foods, *Br. J. Nutr.*, 46, 403, 1981.
2. **Burr, M. L. and Merrett, T. G.,** Food intolerance: a community survey, *Br. J. Nutr.*, 49, 217, 1983.
3. **Anderson, J. A.,** Non-immunologically-mediated food sensitivity, *Nutr. Rev.*, 42, 109, 1984.
4. **Sampson, H. R.,** IgE-mediated food intolerance, *J. Allergy Clin. Immunol.*, 81, 495, 1988.
5. **Buissert, P. D., Heinzelmann, D. I., Youlten, L. J. F., and Lessof, M. H.,** Prostaglandin-synthesis inhibitors in prophylaxis of food intolerance, *Lancet*, 1, 906, 1978.
6. **Leavitt, M. D.,** Intestinal gas, *Proc. Nutr. Soc.*, 44, 165, 1985.
7. **Truswell, A. S.,** Food sensitivity, *Br. Med. J.*, 291, 951, 1985.
8. **Kao, H. W. and Zipser, R. D.,** Exaggerated prostaglandin production by colonic smooth muscle in rabbit colitis, *Dig. Dis. Sci.*, 33, 697, 1988.
9. **Sanders, K. M.,** Role of prostaglandins in regulating gastric motility, *Am. J. Physiol.*, 247, G117, 1984.
10. **Tollstrom, T., Hellstrom, P. M., Johansson, C., and Pernow, B.,** Effects of prostaglandins E_2 and $F_{2\alpha}$ on motility of small intestine in man, *Dig. Dis. Sci.*, 33, 552, 1988.
11. **Isselbacher, K. J.,** The role of arachidonic acid metabolites in gastrointestinal homeostasis — biochemical, histological and clinical gastrointestinal effects, *Drugs*, 33, 38, 1987.

12. **Jones, V. A., Shorthouse, M., McLaughlan, P., Workman, E., and Hunter, J. O.,** Food intolerance: a major factor in the pathogenesis of irritable bowel syndrome, *Lancet*, 2, 1115, 1982.
13. **Laursen, L. S., Naesdal, J., Bukhave, K., Lauritsen, K., and Rask-Madsen, J.,** Selective 5-lipoxygenase inhibition in ulcerative colitis, *Lancet*, 335, 683, 1990.
14. **Radimirov, R. and Venkova, K.,** Effects of low extracellular Ca^{2+} on prostaglandins $F_{2\alpha}$ and E_2 action in longitudinal muscle of guinea-pig caecum, *Gen. Pharmacol.*, 19, 86, 1988.
15. **Kanof, P. D., Johns, C., Davidson, M., Siever, L. J., Coccaro, E. F., and Davis, K. L.,** Prostaglandin receptor sensitivity in psychiatric disorders, *Arch. Gen. Psychiatry*, 43, 987, 1986.
16. **Sadoshima, J.-I., Akaike, N., Kanaide, H., and Nakamura, M.,** Cyclic AMP modulates Ca-activated K channel in cultured smooth muscle cells of rat aortas, *Am. J. Physiol.*, 255, H754, 1988.
17. **Narducci, F., Bassotti, G., Granata, M. T., Pelli, M. A., Gaburri, M., Palumbo, R., and Morelli, A.,** Colonic motility and gastric emptying in patients with irritable bowel syndrome, *Dig. Dis. Sci.*, 31, 241, 1986.
18. **Rask-Madsen, J., Bukhave, K., and Beubler, E.,** Influence on intestinal secretion of eicosanoids, *J. Intern. Med.*, 228, (Suppl. 1), 137, 1990.
19. **Tseng, C. F., Mikajiri, A., Shibuya, M., Goda, Y., Elizuka, Y., Padmawinata, K., and Sankawa, U.,** Effects of some phenolics on the prostaglandin synthesizing enzyme system, *Chem. Pharmacol. Bull.*, 34, 1380, 1986.
20. **Ribereau-Gayon, P.,** *Plant Phenolics*, Hafner Publishing, New York, 1972, chap. 1.
21. **Robinson, T.,** *The Organic Constitutents of Higher Plants — Their Chemistry and Interrelationships*, 4th ed., Cordus Press, North Amherst, MA, 1980, chap. 4.
22. **Blalock, J. E., Archer, D. L., and Johnson, H. M.,** Anticellular and immunosuppressive activities of foodborne phenolic compounds, *Proc. Soc. Exp. Biol. Med.*, 167, 391, 1981.
23. **Stanley, R. G. and Linskens, H. F.,** *Pollen — Biology, Biochemistry, Management*, Springer-Verlag, New York, 1974, chap. 15.
24. **Salano, A. R., Dada, L. A., Sardañons, M. L., Sánchez, M. L., and Podestá, E. J.,** Leukotrienes as common intermediates in the cAMP dependent and independent pathways in adrenal steroidogenesis, *J. Steroid Biochem.*, 27, 745, 1987.
25. **Gardner, R. W., McGovern, J. J., and Brenneman, L. D.,** The role of plant and animal phenyls in food allergy, paper presented at 37th Annu. Congr. American College of Allergists, Washington, D.C. April 4-8, 1981. *(See appendix.)*
26. **Archer, D. L.,** Immunotoxicology of foodborne substances: an overview, *J. Food Prot.*, 41, 983, 1978.
27. **Crosby, D. G.,** Natural toxic background in the food of man and his animals, *J. Agric. Food Chem.*, 17, 532, 1969.
28. **Singleton, V. L. and Kratzer, F. H.,** Plant phenolics, in *Toxicants Occurring Naturally in Foodstuffs*, 2nd ed., National Academy of Sciences, Washington, D.C., 1973, 309.
29. **Anderson, R. L., Rackis, J. J., and Tallent, W. H.,** Biologically active substances in soy products, in *Soy Protein and Human Nutrition*, Wilcke, H. L., Hopkins, D. T., and Waggle, D. H., Eds., Academic Press, New York, 1979, 209.

30. **Udenfriend, S., Lovenberg, W., and Sjoerdsman, A.,** Physiologically active amines in common fruits and vegetables, *Arch. Biochem. Biophys.*, 85, 187, 1959.
31. **Edwards, S. T. and Sandine, W. E.,** Public health significance of amines in cheese, *J. Dairy Sci.*, 64, 2431, 1981.
32. **Smith, T. A.,** Amines in food, *Food Chem.*, 6, 169, 1980-81.
33. **Waalkes, T. P., Sjoerdma, A., Craveling, C. R., Weissbach, H., and Udenfriend, S.,** Serotonin, norepinephrine and related compounds in bananas, *Science*, 127, 648, 1958.
34. **Ormsbee, H. S., III and Fondacaro, J. D.,** Action of serotonin on the gastrointestinal tract, *Proc. Soc. Exp. Biol. Med.*, 178, 333, 1985.
35. **Barnes, J. M., et al.,** Topographical distribution of $5HT_3$ recognition sites in the ferret brain stem, *Naunyn-Schmiedebergs Arch. Pharmacol.*, 342, 17, 1990.
36. **Engelman, K., Lovenberg, W., and Sjoerdsman, A.,** Inhibition of serotonin synthesis by para-chlorophenylalanine in patients with carcinoid syndrome, *N. Engl. J. Med.*, 277, 1103, 1967.
37. **Szasz, G.,** Cholinergic agonists, in *Pharmaceutical Chemistry of Adrenergic and Cholinergic Drugs*, CRC Press, Boca Raton, FL, 1985, chap. 3.
38. **Orgell, W. H.,** Inhibition of human plasma cholinesterase *in vitro* by alkaloids, glycosides, and other natural substances, *Lloydia*, 26, 36, 1963.
39. **Endo, M. and Kitazawa, T.,** Calcium ions and contraction of smooth muscle by alpha-adrenergic stimulation, in *Calcium Entry Blockers and Tissue Protection*, Godfraind, T., Vanhoutte, P. M., Govoni, S., and Paoletti, R., Eds., Raven Press, New York, 1985, 81.
40. **Vergroesen, A. J.,** Hypothesis for interactions between acetylcholine and prostaglandin biosynthesis: an introduction, in *Nutrition and the Brain*, Vol. 5, Barbeau, A., Crowden, J. H., and Wurtman, J. R., Eds., Raven Press, New York, 1979.
41. *Merck Index*, 10th ed., Merck and Co., Rahway, NJ, 1983.
42. **Whitaker, J. R. and Feeney, R. E.,** Enzyme inhibitors in food, in *Toxicants Occurring Naturally in Foods*, 2nd ed., National Academy of Sciences, Washington, D.C., 1973, 276.
43. **Osborne, F. G., Willis, K. H., and Barrett, G. M.,** Survival of bromate and bromide in bread baked from white flour containing potassium bromate, *J. Sci. Food Agric.*, 42, 255, 1988.
44. **Mathias, J. R., Ferguson, K. L., and Clench, M. H.,** Debilitating "functional" bowel disease controlled by leuprolide acetate, a gonadotropin-releasing hormone (GnRH) analog, *Dig. Dis. Sci.*, 34, 761, 1989.
45. **Nes, W. R. and McKean, M. L.,** *Biochemistry of Steroids and Other Isopentenoids*, University Park Press, Baltimore, MD, 1977, chap. 10.
46. **Smith, A. L.,** *Principles of Microbiology*, 10th ed., C. V. Mosby, St. Louis, MO, 1985, 547.
47. **Alhadeff, L., Gualtieri, C. T., and Lipton, M.,** Toxic effects of water-soluble vitamins, *Nutr. Rev.*, 42, 33, 1984.
48. **Lewin, S.,** *Vitamin C: Its Molecular Biology and Medical Potential*, Academic Press, New York, 1976.

3

Respiratory Inhibitions

I. HAY FEVER

Allergic rhinitis is one of the most familiar forms of allergic responses, particularly during the pollen seasons. Traditionally, IgE, with associated histamine, is cited as the principal activating factor in allergic reactions such as hay fever. However, Shumacher and Pain[1] found no correlation between total IgE levels and nasal sensitivity. More attention is now being given to other inflammatory compounds such as the eicosanoids.[2-4] For instance, the slow-reacting substance of anaphylaxis (composed of leukotrienes C, D, and E) is released *in vitro* by the interaction of antigen and IgE antibody on human mast cells and basophils.[5] Intranasal challenge of ragweed-sensitive patients with pollen grains was significantly correlated with the release of the peptide leukotrienes from nasal cells in a study by Creticos et al.[5] Nonallergic subjects exhibited neither symptoms nor leukotriene release. Increased mucous secretion and sneezing were associated with the allergic response. Of 16 patients who sneezed, 12 released histamine, 12 prostaglandin D_2 (PGD_2), and 9 leukotrienes, whereas none of the control group exhibited such responses. Use of an oral leukotriene antagonist by Fuller et al.[6] reduced the early response to inhaled antigen. This research team suggested further studies with the antagonists to fully elucidate the role of leukotrienes in antigen-induced responses in man.

Holgate and Howarth[7] suggest that edema of the mucous membrane and nasal obstruction could be due to an interaction of such mediators as PGD_2 and leukotriene C_4 (LTC_4) since PGD_2 is a potent dilator, causing vascular engorgement of the mucosa, while LTC_4 increases capillary permeability, causing exudation. The arachidonic acid metabolites (eicosanoids) affect mucous and water content of airway secretions as well as ciliary action.[3] Chloride ion secretions are altered, and there is a change in the status of sodium ions.

Premedication with aspirin did not affect the amount of sneezing or the levels of either histamine or leukotrienes in allergic individuals exposed to ragweed and grass-mix extract.[8] Aspirin did significantly inhibit increases in the cyclooxygenase metabolites PGE_1, PGD_2, PGF_2, 6-keto-PGF_1, and thromboxane. Brown et al.[8] concluded that the early phase of allergic rhinitis is a mast cell-dominated event.

Ragweed pollen *(Parthenium hysterophorus)* is recognized as a very irritating pollen by many susceptible "hay fever" patients.[5] Pollens contain many phenolic compounds which are possibly the agents which activate the production of eicosanoids, resulting in an inflammatory response. Several phenolic compounds have been identified in ragweed pollen, namely caffeic, vanillic, *p*-coumaric, chlorogenic, and ferulic acids.[9] Flavonoids isolated from pollen of *Petunia hybrida* include taxifolin, quercetin, and kaempferol.[10] Flavonoids occur in pollen of many angiosperm and gymnosperm species.[11]

Cross-reactivity among pollens has raised the question as to chemical entities common to those pollens.[12,13] A further cross-reactivity between pollens and foods has increased interest in this aspect of allergic reactions. Dreborg and Foucard[14] were puzzled as to why such botanically unrelated foods as apple, carrot, and potato should share allergens with each other and with birch pollen. Björkstén[15] reported a close clustering of birch pollen with celery, apple, and carrot sensitivity. Another association has been reported between ragweed pollinosis and clinical sensitivity to melon and banana.[16] Symptoms of respiratory allergies have been found to be worse in subjects with coexisting food sensitization.[17]

Mere inhalation of certain food odors can lead to rhinitis. For instance, Pearson[18] described 29 female subjects who experienced sneezing and/or wheezing while they were scraping uncooked potatoes. Castells et al.[19] observed that an 11-year-old girl developed respiratory and other systemic symptoms on contact with potatoes, ingestion of potatoes, and exposure to cooking potatoes or potato pollen. One of the aromatic compounds possibly responsible for the inhalant response is vanillin, which is in potato parings.[20] Aromatic (phenolic) agents used in perfumes, in tobacco, and as food flavoring agents cause adverse respiratory effects in susceptible individuals. Groups of rats were exposed to a submicron aerosol of eugenol for 4 hours.[21] Clinical signs observed during exposure consisted principally of moderately increased salivation, restlessness (indicative of irritation), and abnormal breathing patterns. Addition of cloves (a source of eugenol) to tobacco might well be a source of nasal irritation to both the smoker and those around him.

Becker et al.[22] described the isolation of antigens from cocoa powder, ground coffee, and ragweed pollen which would immunologically cross-react with a glycoprotein isolated from cured tobacco leaves and from cigarette smoke condensate. These antigens contained similar polyphenol haptens and were capable of activating factor XII (Hageman factor)-dependent pathways. The polyphenol haptens included either rutin or quercetin-like groups. Quercetin, or its derivatives, is the most common flavonoid in pollens.[11] The foodstuff containing the highest known content of quercetin is buckwheat seed (2.0 to 2.5%).[11] This commonality in antigenicity could be due to the phenolic compounds which are common to numerous pollens as well as foodstuffs. The mechanism of the adverse effects of the phenolics is described elsewhere (Chapter 1).

II. ASTHMA

Scadding[23] observed that patients with IgE-mediated hypersensitivity reactions are a minority, whereas bronchoconstrictor responses to many chemicals and physical stimuli are the major clinical problems. LTC_4 and LTD_4 have been identified as unique bronchoconstrictors with a possible role in the pathogenesis of asthma.[24] Mortagy et al.[25] suggested that "bronchial irritability syndrome" is a clearer diagnostic term than asthma because of symptoms being provoked by physical or chemical factors in air.

Histamine has long been the mediator identified with clinical asthma. Now the listing also includes LTC_4 and LTD_4^2 as well as LTB_4, LTE_4, PGD_2, $PGF_{2\alpha}$, thromboxane, platelet-activating factor, neutrophil chemotactic factor, and eosinophil chemotactic factor of anaphylaxis.[26-28] There are reports that LTC_4 is 3800 times as active as histamine in decreasing the airflow in lungs of normal human volunteers who inhale the agents.[29] *In vivo* release of leukotrienes C, D, and E after exposure to ragweed pollen in patients sensitive to ragweed resulted in clinical responses typical of "hay fever".[30] Nonallergic subjects had neither symptoms nor leukotriene release.

An interesting hypothesis offered is that substances with a benzene ring circulate through the lungs and by an unknown mechanism induce lung irritation, manifested as bronchospasm.[31] The lungs may be exposed to the benzene ring via pollens, perfumes, and air pollutants.[32] In addition, some estimate that approximately 50% of the benzene ring structures absorbed from the gastrointestinal tract are excreted via the lungs.[33] Bronchoconstriction due to tartrazine and other dyes was reported in 2% of patients by Weber et al.[34]

Kay[35] prepared an instructive scheme (Figure 3.1) to illustrate the relationships between early, late-phase, and ongoing chronic asthma. Vacillations in forced expiratory volume (FEV_1; Figure 3.1) in association with chemical activating factors, over time, are informative. Platelet activating factor (PAF), PGD_2, histamine, LTC_4, and LTD_4 are released from mediator cells (MC) in response to antigen in the early phase (Figure 3.1) The early allergic asthmatic reaction is predominately mast cell mediated, elaborating LTC_4, PGD_2, various chemotactic peptides, proteolytic enzymes, and proteoglycans.[35] Mast cells are involved in immediate bronchoconstriction due to both immunologic and nonimmunologic stimuli.[35,36]

Aspirin has been used to relieve bronchospasms in some patients, but has also been implicated as inducing bronchospasms in other patients.[31] Collier[37] concluded that the balance between prostaglandins of the F series, which are bronchoconstrictors, may be disturbed in patients with

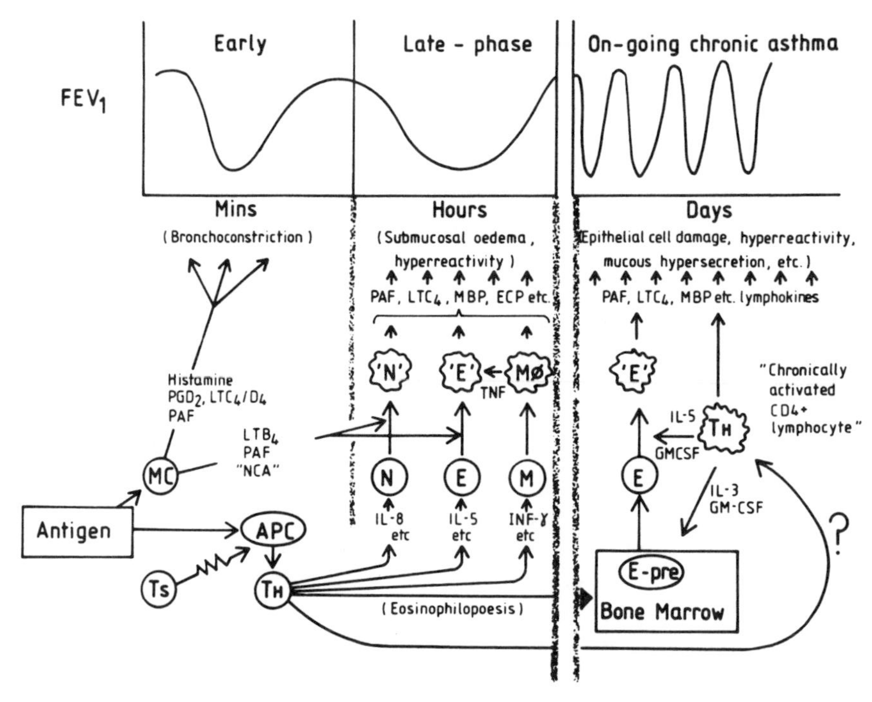

Figure 3.1 A scheme to explain the relationship between early, late-phase, and ongoing chronic asthma. The early phase is believed to be dependent on products derived from Fc receptor-bearing mediator cells *(MC)*. It is suggested that the late-phase reaction is dependent on mast cell-derived chemotactic factors (i.e., LTB_4, platelet activating factor [PAF], and NCA) as well as products derived from CD4 helper cells. It is speculated that an important feature of ongoing chronic asthma is "chronically activated CD_4 lymphocytes" that orchestrate eosinophil maturation and activation which in turn lead to epithelial cell damage, hyperreactivity, and mucus hypersecretion. (The diagram is greatly simplified; for instance, interleukin-1 and interleukin-8 are derived from T-cells as well as antigen-presenting cells.) (From Kay, A. B., *J. Allergy Clin. Immunol.*, 87, 893, 1991. With permission.)

asthma. Asthmatics are abnormally sensitive to the bronchoconstrictor effects of prostaglandin $F_{2\alpha}$. Since aspirin blocks the cyclooxygenase pathway, it is likely that arachidonic acid is diverted to the lipoxygenase pathway, yielding leukotrienes.[37-43] This could be the reason for the adverse response to aspirin in some patients, since LTC_4 and LTD_4 have been shown to be potent bronchoconstricting agents both in animals and in normal and asthmatic humans.[23,42,43] Christie et al.[42] have verified the hypothesis that cyclooxygenase inhibitors such as aspirin and indomethacin may divert arachidonic acid to lipoxygenase products, particularly in aspirin-sensitive patients. They did this by measuring urinary LTE_4 as a marker since it is derived enzymatically from LTD_4, which in turn is derived from LTC_4. Urine of aspirin-sensitive subjects with asthma contained significantly higher levels of LTE_4 than subjects with asthma who were not aspirin intolerant.

Patients with asthma are more responsive to LTC_4 and LTD_4 than normal subjects.[23,45,46] When comparing LTD_4 with histamine, Griffen and co-workers[44] learned that LTD_4 was 140 times more potent (on a molar basis) than histamines or the prostaglandins in its effect on the pulmonary airways.

A study was designed to identify the mechanism of LTD_4-induced bronchoconstriction in normal subjects. The investigators concluded that, at least in part, the bronchoconstriction was a direct effect of LTD_4 on airway smooth muscle.[45] Lung tissue of two birch pollen-sensitive asthmatics released LTC_4, LTD_4, and LTE_4 upon challenge with the pollen.[27] Hyperventilation with cold, dry air also causes asthmatic bronchoconstriction in asthmatics, and inhibition of the 5-lipoxygenase enzyme was associated with a significant amelioration of this asthmatic response.[46] Evidence of the release of LTC_4 by eosinophils in asthmatics prompted Bruynzeel and Verhagen[30] to suggest that future asthma therapy should be focused on preventing leukotriene formation by these cells. Chemicals which are efficacious in that role will be discussed in Chapters 14 and 15.

Thromboxane generation from inflammatory cells (e.g., neutrophils) resulting in airway hyperresponsiveness in allergic dogs exposed to ragweed has been reported.[47]

Patients with asthma are also hyperresponsive to methacholine (acetyl-β-methylcholine).[24] In fact, inhalation of increasing levels of methacholine produced increasing bronchoconstriction in both patients with asthma and controls in a study reported by Adelroth et al.[24]

Peak expiratory flow rates decreased linearly with increasing formaldehyde (HCHO) exposure in children.[48] There were significantly greater prevalence rates of asthma and chronic bronchitis in children from houses with HCHO levels of 60 to 120 ppb than in those less exposed, especially in children also exposed to environmental tobacco smoke. The effects of HCHO in adults were much less evident than in children.[48]

Sulfites are used as preservatives to inhibit nonenzymic browning of foodstuffs. They have been shown to provoke asthma in susceptible individuals.[49,50] Allen[49] has promoted the hypothesis that, since there is a similarity of reactions by sensitive individuals to inhaled gaseous sulfur dioxide (SO_2) and to ingested sulfites, the asthma response following the ingestion of acidified solutions of sulfite is due to supersensitivity to SO_2 inhaled during swallowing. Bronchoconstriction after inhalation of low concentrations of SO_2 is a feature of asthma.[50] A cholinergic reflex response has been identified upon inhalation of SO_2 and bronchoconstriction to inhaled SO_2 and has been blocked by cutting vagal trunks in animals as well as by using nebulized atropine.[50] A postulate is that the mechanism of sulfite-induced bronchoconstriction is by sulfonation of cholinergic receptors by the sulfite ions.[51]

Stable inorganic sulfites have also been found in plumes and effluents from power plants and smelters.[52] Chen et al.[52] reported decreased lung capacity in guinea pigs exposed to submicrometer sodium sulfite aerosols. They proposed that the inflammatory response detected in the lungs of the guinea pigs may have been caused by alteration of the pH of the airways upon exposure to alkaline sodium sulfite aerosols rather than to sulfite itself.

Tachykinins have been shown to affect the tracheobronchial tree directly and to interact with other bronchoconstrictive mediators.[53] Treatment with the phenolic compound capsaicin reportedly depletes the sensory C-fibers of their tachykinin content.[53] Alving et al.[53] challenged rats with capsaicin, and the results demonstrated a link between tachykinins, serotonin, the immune system, and clinical lung reactivity. The bronchoconstrictor effect of capsaicin has also been identified in humans.[54]

The magnesium ion (Mg^{2+}) has a bronchodilating effect in bronchial asthma.[55] Those identifying this effect used literature citations for possible mechanisms of Mg^{2+} action, which included an inhibitory action on (1) smooth muscle contraction, (2) histamine release from mast cells, and (3) acetylcholine release from nerve terminals. Mg^{2+} appears to counter Ca^{2+} effects on smooth muscles via activation of cAMP. This is because Mg^{2+} is a cofactor in the formation of cyclic guanosine monophosphate (cGMP), which activates relaxation of smooth muscles.[56] In addition, Mg^{2+} inhibits slow inward Ca^{2+} current and Ca^{2+}-induced Ca^{2+} release.[55] In fact, lower Mg uptake or Mg deficiency may play a role in some types of asthma.[55] Intravenous as well as oral Mg administration may be useful as a supporting therapy in controlling bronchial asthma.

REFERENCES

1. **Shumacher, M. J. and Pain, M. C. F.,** Nasal challenge testing in grass pollen hay fever, *J. Allergy Clin. Immunol.*, 64, 202, 1979.
2. **Togias, A., Naclerio, R. M., Proud, D., Pipkorn, U., Bascom, R., Iliopoulos, O., Kagey-Sobotka, A., Norman, P. S., and Lichstenstein, L. M.,** Studies on the allergic and nonallergic nasal inflammation, *J. Allergy Clin. Immunol.*, 81, 782, 1988.
3. **Henke, D., Danilowicz, R. M., Curtis, J. F., Boucher, R. C., and Eling, T. E.,** Metabolism of arachidonic acid by human nasal and bronchial epithelial cells, *Arch. Biochem. Biophys.*, 267, 426, 1988.
4. **Barnes, N. C. and Costello, J. F.,** Airway hyperresponsiveness and inflammation, *Br. Med. Bull.*, 43, 445, 1987.
5. **Creticos, P. S., Peters, S. P., Adkinson, N. F., Jr., Naclerio, R. M., Hayes, E. C., Norman, P. S., and Lichtenstein, L. M.,** Peptide leukotriene release after antigen challenge in patients sensitive to ragweed, *N. Engl. J. Med.*, 310, 1626, 1984.

6. **Fuller, R. W., Black, P. N., and Dollery, C. T.,** Effect of the oral leukotriene D_4 antagonist LY171883 on inhaled and intradermal challenge with antigen and leukotriene D_4 in atopic subjects, *J. Allergy Clin. Immunol.*, 83, 939, 1989.
7. **Holgate, S. T. and Howarth, P. H.,** What's new about hay fever?, *Br. Med. J.*, 291, 1, 1985.
8. **Brown, M. S., Peters, S. P., Adkinson, F., Jr., Proud, D., Kagey-Sobotka, A., Norman, P. S., Lichtenstein, L. M., and Neclerio, R. M.,** Arachidonic acid metabolites during nasal challenge, *Arch. Otolaryngol. Neck Surg.*, 113, 179, 1987.
9. **Mersie, W. and Singh, M.,** Effects of phenolic acids and ragweed parthenium *(Parthenium hysterophorus)* extracts on tomato growth and nutrient and chlorophyll content, *Weed Sci.*, 36, 278, 1988.
10. **Zerback, R., Bokel, M., Geiger, H., and Hess, D.,** A kaempferol 3-glucosylgalactoside and further flavonoids from pollens of *Petunia hybridia, Phytochemistry*, 28, 897, 1989.
11. **Stanley, R. G. and Linskens, H. F.,** *Pollen — Biology, Biochemistry, Management*, Springer-Verlag, New York, 1974, chap. 15.
12. **Weber, R. W.,** Cross-reactivity among pollens, *Ann. Allergy*, 46, 208, 1981.
13. **Gonzáles, R. M., Cortés, C., Conde, J., Negro, J. M., Rodriguez, J., Tursi, A., Wuthrich, B., and Carreira, J.,** Cross-reactivity among five major pollen allergens, *Ann. Allergy*, 59, 149, 1987.
14. **Dreborg, S. and Foucard, T.,** Allergy to apple, carrot and potato in children with birch pollen allergy, *Allergy*, 38, 167, 1983.
15. **Björksten, F.,** Food sensitivity — Marabou Symposium, *Nutr. Rev.*, 42, 131, 1984.
16. **Anderson, L. B., Jr., Frefuss, E. M., Logan, J., Johnstone, D. E., and Glaser, J.,** Melon and banana sensitivity coincident with ragweed pollinosis, *J. Allergy*, 45, 310, 1970.
17. **Fiorini, G., et al.,** Symptoms of respiratory allergies are worse in subjects with coexisting food sensitization, *Clin. Exp. Allergy*, 20, 689, 1990.
18. **Pearson, R. S.,** Potato sensitivity, an occupational allergy in housewives, *Acta Allergol.*, 21, 507, 1966.
19. **Castells, M. C., Pascual, C., Esteban, M., and Ojeda, J. A.,** Allergy to white potato, *J. Allergy Clin. Immunol.*, 78, 1110, 1986.
20. **Merck Index,** 10th ed., Merck & Co., Rahway, NJ, 1983.
21. **Clark, G. C.,** Acute inhalation toxicity of eugenol in rats, *Arch. Toxicol.*, 62, 381, 1988.
22. **Becker, C. G., Hamont, N. V., and Wagner, M.,** Tobacco, cocoa, coffee and ragweed: cross-reacting allergens that activate factor-XII-dependent pathways, *Blood*, 58, 861, 1981.
23. **Scadding, J. G.,** Asthma and bronchial reactivity, *Br. Med. J.*, 294, 1115, 1987.
24. **Adelroth, E., Morris, M. M., Hargreave, F. E., and O'Byrne, P. M.,** Airway responsiveness to leukotrienes C_4 and D_4 and to methacholine in patients with asthma and normal controls, *N. Engl. J. Med.*, 315, 480, 1986.
25. **Mortagy, A. K., Howell, J. B. L., and Water, W. E.,** Respiratory symptoms and bronchial reactivity: identification of a syndrome and its relation to asthma, *Br. Med. J.*, 293, 525, 1986.
26. **Townley, R. J. and Hopp, R. J.,** Inhalation methods for the study of airway responsiveness, *J. Allergy Clin. Immunol.*, 80, 111, 1987.

27. **Marx, J. L.,** The leukotrienes in allergy and inflammation, *Science,* 215, 1380, 1982.
28. **Creticos, P. S., Peters, S. P., Adkinson, N. F., Jr., Naclerio, R. M., Hayes, E. C., Norman, P. S., and Lichtenstein, L. M.,** Peptide leukotriene release after antigen challenge in patients sensitive to ragweed, *N. Engl. J. Med.,* 310, 1626, 1984.
29. **Dahlén, S. -E., Hansson, G., Hedqvist, P., Bjorck, T., Granström, E., and Dahlén, B.,** Allergen challenge of lung tissue from asthmatics elicits bronchial contraction that correlates with the release of leukotrienes C_4, D_4, and E_4, *Proc. Natl. Acad. Sci. U.S.A.* 80, 1712, 1983.
30. **Bruynzeel, P. L. B. and Verhagen, J.,** The possible role of particular leukotrienes in the allergen-induced late-phase asthmatic reaction, *Clin. Exp. Allergy,* 19, 25, 1989.
31. **Kounis, N. G.,** A review: drug-induced bronchospasm, *Ann. Allergy,* 37, 285, 1976.
32. **Randolph, T. G.,** *Human Ecology and Susceptibility to the Chemical Environment,* Charles C Thomas, Springfield, IL, 1972, 106.
33. **Goodman, L. S. and Gilman, A.,** *The Pharmacological Basis of Therapeutics,* 5th ed., Macmillan, New York, 1979, chap. 1.
34. **Weber, R. W., Hoffman, M., and Raine, D. A.,** Incidence of bronchoconstriction due to aspirin, azo dyes, non-azo dyes and preservatives in a population of perennial asthmatics, *J. Allergy Clin. Immunol.,* 64, 33, 1979.
35. **Kay, A. B.,** Asthma and inflammation, *J. Allergy Clin. Immunol.,* 87, 893, 1991.
36. **Wasserman, S. I.,** Mast cell-mediated inflammation in asthma, *Ann. Allergy,* 63, 547, 1989.
37. **Collier, H. O.,** Prostaglandins and aspirin, *Nature,* 232, 17, 1971.
38. **Szczeklik, A.,** A cyclooxygenase theory of aspirin-induced asthma, *Eur. Respir. J.,* 3, 588, 1990.
39. **Vane, J. and Botting, R.,** Inflammation and the mechanism of action of anti-inflammatory drugs, *FASEB J.,* 1, 89, 1987.
40. **Stevenson, D. D. and Lewis, R. A.,** Proposed mechanisms of aspirin sensitivity reactions (editorial), *J. Allergy Clin. Immunol.,* 80, 788, 1987.
41. **Bosso, J. V., Schwartz, L. B., and Stevenson, D. D.,** Tryptase and histamine release during aspirin-induced respiratory reactions, *J. Allergy Clin. Immunol.,* 88, 830, 1991.
42. **Christie, P. E., Tagari, P., Ford-Hutchinson, A. W., Charlesson, S., Chee, P., Arm, J. P., and Lee, T. H.,** Urinary LTE_4 concentrations increase after aspirin challenge in aspirin-sensitive asthmatic subjects, *Am. Rev. Respir. Dis.,* 143, 1025, 1991.
43. **Lee, T. H., Smith C. M., Arm, J. P., and Christie, P. E.,** Mediator release in aspirin-induced reactions (editorial), *J. Allergy Clin. Immunol.,* 88, 827, 1991.
44. **Griffen, M., Weiss, J. W., Leitch, A. G., McFadden, E. R., Jr., Corey, E. J., Austin, K. F., and Drazen, J. M.,** Effects of leukotriene D on the airways in asthma, *N. Engl. J. Med.,* 308, 436, 1983.
45. **Smith, L. J., Kern, R., Patterson, R., Krell, R., Bernstein, P. R.,** Mechanisms of leukotriene D_4-induced bronchoconstriction in normal subjects, *J. Allergy Clin. Immunol.,* 80, 340, 1987.
46. **Israel, E., Dermarkarian, R., Rosenberg, M., Sperling, R., Taylor, G., Rubin, P., and Drazen, J. M.,** The effects of a 5-lipoxygenase inhibitor on asthma induced by cold, dry air, *N. Engl. J. Med.,* 323, 1740, 1990.

47. **Chung, K. F., Aizawa, H., Becker, A. B., Frick, O., Gold, W. M., and Nadel, J. A.,** Inhibition of antigen-induced airway hyper-responsiveness by a thromboxane synthetase inhibitor (OKY-046) in allergic dogs, *Am. Rev. Respir. Dis.*, 134, 258, 1986.
48. **Krzyzanowski, M., Quackenboss, J. J., and Lebowitz, M. D.,** Chronic respiratory effects of indoor formaldehyde exposure, *Environ. Res.*, 52, 117, 1990.
49. **Allen, K. H.,** Asthma induced by sulphites, *Food Technol. Aust.*, 37, 506, 1985.
50. **Bush, R. K., Taylor, S. L., and Buss, W.,** A critical evaluation of clinical trials in reactions to sulfites, *J. Allergy Clin. Immunol.*, 78, 191, 1986.
51. **Stenacker, A.,** A suggested mechanism of action for sulfite sensitivity, *J. Allergy Clin. Immunol.*, 77, 116, 1985.
52. **Chen, L. C., Lam, H. F., Ainsworth, D., Gutty, J., and Amdur, M. O.,** Functional changes in the lungs of guinea pigs exposed to sodium sulfite aerosols, *Toxicol. Appl. Pharmacol.*, 89, 1, 1987.
53. **Alving, K., Ulfgren, A. -K., Lundberg, J. M., and Ahlstedt, S.,** Effect of capsaicin on bronchial reactivity and inflammation in sensitized adult rats, *Int. Arch. Allergy Appl. Immunol.*, 82, 377, 1987.
54. **Fuller, R. W., Dixon, C. M. S., and Barnes, P. J.,** Bronchoconstrictor response to inhaled capsaicin in humans, *J. Appl. Physiol.*, 58, 1080, 1985.
55. **Okayama, H., Aikawa, T., Okayama, M., Sasaki, H., Mue, S., and Takishima, T.,** Bronchodilating effect of intravenous magnesium sulfate in bronchial asthma, *J.A.M.A.*, 257, 1076, 1987.
56. **Waldman, S. A. and Murad, F.,** Cyclic GMP synthesis and function, *Pharmacol. Rev.*, 39, 163, 1987.

4

Arthritis

An estimated 1 to 2% of the world's population suffers from rheumatoid arthritis, the most severe of arthritic diseases.[1] In 1981, the Arthritis Foundation reported that 31 million Americans were affected by arthritis and that an estimated 6.5 million Americans suffered from rheumatoid arthritis.[2] Other forms of arthritis include osteoarthritis, ankylosing spondylitis, systemic lupus erythematosus, and gout.

Rheumatoid arthritis is primarily a disease of the synovial joints. Afflicted joints are swollen and painful, and destruction of articular tissues (cartilage, bone, etc.) is found in a proportion of patients.[1] Zvaifler[3] described the inflammatory reaction effect on the synovial lining of arthritic joints as appearing edematous and the synovium protruding into the joint cavity as slender villous projections. He further observed that synovial lining cells were hypertrophied and reached to a depth of six to ten cells, whereas normally there are only one to three cell layers. Venous distention results from swollen endothelial cells, and capillary obstruction is common.[3]

Mediators of rheumatoid arthritis have been identified as prostaglandins, leukotrienes, and interleukin-1.[1,4-6] There is no evidence to suggest that histamine plays a significant role in the pathology of this disorder.[1] Local synthesis of prostaglandins in the rheumatoid joint apparently sensitizes pain receptors to chemical stimuli.[1] Relief of pain in the rheumatoid joint by the use of aspirin gives validity to the hypothesis that cyclooxygenase products (prostaglandins, particularly PGE) are responsible for the pain.[1,4] A limitation in the use of aspirin is sensitivity reactions in certain individuals.[7] Corticosteroids have also been used in small amounts to treat arthritic patients since they interfere with the release of prostaglandins.[3] Secretory phospholipase A_2 has been purified from synovial fluid from patients with rheumatoid arthritis.[8-10] The role of this enzyme group in catalyzing the hydrolysis of the 2-acyl bond of phospholipids with release of arachidonic acid and subsequent formation of eicosanoids was discussed in detail in Chapter 1. Chemicals found to be effective in inhibiting the action of phospholipase A_2 are listed in Chapters 14 and 15.

LTB_4 has been detected in the synovial fluid and tissues of patients with acute rheumatoid arthritis[11] as well as patients with gouty effusions.[12] The concentration of a PGE_2-like substance is about 20 μg/l in the synovial fluid of patients with rheumatoid arthritis.[13]

Eicosapentaenoic acid (EPA) from fish oils has an anti-inflammatory effect in rheumatoid arthritis.[14] Reduction in joint swelling associated with *Salmonella*-induced arthritis in rats has been attributed to EPA supplementation of diets.[15] In a review paper, Moncada and Salmon[16] cited possible reasons for the anti-inflammatory effects of EPA. One reason proposed is the direct competition of the conversion of arachidonic acid (AA) by the cyclooxygenase enzyme. EPA, unlike AA, is a poor substrate for platelet, blood vessel, kidney, and heart cyclooxygenase; therefore, it was suggested that EPA was a competitive inhibitor of AA at the level of the platelet cyclooxygenase, which could be a major site of action.[17]

Gouty arthritis arising from hyperuricemic individuals has been the object of numerous studies in nephropathy.[18] Multiple environmental factors, along with genetic factors, determine plasma uric acid levels.[19] Uric acid possibly exerts its effect by triggering the release of leukotrienes or by activating enzyme systems responsible for the synthesis of leukotrienes. Tuomilehto et al.[19,20] found a strong renal involvement in the balance of plasma uric acid which might have reflected certain dietary patterns, such as a high intake of protein, fats, and certain local vegetables grown in Fiji. They found a significant intercorrelation between plasma uric acid, impaired glucose tolerance, and plasma cholesterol in men but not in women. Hyperglycemia enhances uric acid excretion by possibly impairing tubular reabsorption of uric acid.[21] Common exposure to uric acid via foodstuffs and catabolic products of purines exposes the susceptible individual to an unavoidable crisis. Adding to the stress load is possible exposure to xanthine, theobromine, caffeine, theophylline, and chemically related structures which might activate the same physiological response as uric acid.

Some patients have reported an exacerbation of rheumatoid arthritis in association with the ingestion of certain foods. Parke and Hughes[22] reported verification of this by exclusion of dairy products from the diet of a 38-year-old mother with an 11-year history of progressive, erosive, seronegative rheumatoid arthritis. Elimination of milk, cheese, and butter from her diet for 3 weeks reduced synovitis, morning stiffness, and ring size. Grip strengths, Ritchie index, and visual pain analogue scores were also improved. The authors observed that patients may not relate joint symptoms to an offending food until that antigen is removed from the diet.

The need for more research into the association between food allergies and arthritis was suggested by Moment.[23] He referred to several hypotheses as to specific food groups which might be responsible, but concluded that elimination of such groups has not always been successful.

However, he suggested that "it would be surprising indeed if food allergies were not involved in one or more of the forms of arthritis from gout and lupus on down the line." In support of this was a report that inflammatory arthritis is commonly encountered in patients with inflammatory bowel disease.[24]

Herbs and spices have been prescribed for the relief of arthritis in traditional medicine in India.[25] This fact prompted Srivastava[25] to investigate the effects of onion and ginger on the metabolism of arachidonic acid. Onion tended to slightly increase the amount of thromboxane B_2 (TxB_2) in blood, whereas ginger reduced the TxB_2 level by about 37%. He also obtained evidence that ginger consumption might reduce pain and improve the movement of joints in arthritic patients. He and co-workers hypothesized that ginger might be able to do so by being a dual inhibitor of the cyclooxygenase and lipoxygenase pathways, thus reducing the production of prostaglandins and leukotrienes, respectively. Gingerol is the major phenolic compound and pungent component in ginger oil[26] and may be the principal active agent. This will have to be verified.

Three compounds (caffeic acid, eupatilin, and 4-dimethyleupatilin) isolated from the Chinese plant *Artemisia rubripes nakai* reportedly are selective 5-lipoxygenase inhibitors which do not inhibit prostaglandin synthesis.[27] These compounds may also be functional as pharmacological agents in the prevention and/or treatment of arthritis and gout.

Serotonin and 5-hydroxyindoleacetic acid levels were measured in patients with arthritis and food intolerance.[28] Ingestion of certain foods caused exacerbation of arthritis, with a fall in plasma serotonin levels and a subsequent increase in 5-hydroxyindoleacetic acid. This release of serotonin was assumed by the investigators to enhance the deposition of immune complexes in synovial membranes, thus contributing to the inflammatory process in rheumatoid arthritis.

REFERENCES

1. **Henderson, B., Pettipher, E. R., and Higgs, G. A.,** Mediators of rheumatoid arthritis, *Br. Med. J.*, 43, 415, 1987.
2. **McCormack, P.,** Health Editor, Arthritis probers optimistic, *Provo Herald (Provo, UT)*, p. 46, July 19, 1981.
3. **Zvaifler, N. J.,** Pathogenesis of the joint disease of rheumatoid arthritis, *Am. J. Med.*, 75(6A), 3, 1983.
4. **Lands, W. E. M.,** Control of prostaglandin biosynthesis, *Prog. Lipid Res.*, 20, 875, 1981.
5. **Harris, E. D.,** Pathogenesis of rheumatoid arthritis, *Am. J. Med.*, 80 (Suppl. 4B), 4, 1986.

6. **Seilhammer, J. J., Plant, S., Pruzanski, W., Schilling, J., Stefanski, E., Vadas, P., and Johnson, L. K.,** Multiple forms of phospholipase A_2 in arthritic synovial fluid, *J. Biochem.*, 106, 38, 1989.
7. **Stevenson, D. D. and Lewis, R. A.,** Proposed mechanism of aspirin sensitivity reactions, *J. Allergy Clin. Immunol.*, 80, 788, 1987.
8. **Stefanski, E., Pruzanski, W., Sterby, W., and Vadas, P.,** Purification of a soluble phospholipase A_2 from synovial fluid in rheumatoid arthritis. *J. Biochem. (Tokyo)*, 100, 1297, 1986.
9. **Kramer, R. M., Hession, C., Johansen, B., Hayes, G., McGray, P., Chow, E. P., Tizard, R., and Pepinsky, R. B.,** Structure and properties of a human non-pancreatic phospholipase A_2 from synovial fluid in rheumatoid arthritis, *J. Biol. Chem.*, 264, 5768, 1989.
10. **Wery, J. P.,** Structure of recombinant human rheumatoid arthritic synovial fluid phospholipase A_2 at 2.2 Å resolution, *Nature*, 352, 79, 1991.
11. **Davidson, S. A. and Smith, M. J. H.,** Leukotriene B_4, a mediator of inflammation present in synovial fluid in rheumatoid arthritis, *Ann. Rheum. Dis.*, 42, 677, 1983.
12. **Rae, S. A., Davidson, E. M., and Smith, M. J. H.,** Leukotriene B_4, an inflammatory mediator in gout, *Lancet*, 2, 1122, 1982.
13. **Vane, J.,** The evolution of non-steroidal anti-inflammatory drugs and their mechanism of action, *Drugs*, 33 (Suppl. 1), 18, 1987.
14. **Kremer, J. M., Michalek, A. V., Lininger, L., Huyck, C., Bigauoette, J., Timchalk, M. A., Rynes, R. I., and Zieminski, J.,** Effect of manipulation of dietary fatty acids on clinical manifestations of rheumatoid arthritis, *Lancet*, 1, 184, 1985.
15. **Karmali, R. A., Davies, J., and Volkman, A.,** Role of prostaglandins E_1, E_2, $F_{2\alpha}$, I_2 and thromboxane in *Salmonella*-associated arthritis in rats, *Prostaglandins Leukotrienes Med.*, 8, 437, 1982.
16. **Moncada, S. and Salmon, J. A.,** Leucocytes and tissue injury: the use of eicosapentaenoic acid in the control of white cell activation, *Prog. Lipid Res.*, 25, 563, 1986.
17. **Higgs, E. A., Moncada, S., and Vane, J. R.,** Prostaglandins and thromboxanes from fatty acids, *Prog. Lipid Res.*, 25, 5, 1986.
18. **Gutman, A. B., Ed.,** *Gout: A Clinical Comprehensive*, Burroughs Wellcome, Research Triangle Park, NC, 1971.
19. **Tuomilehto, J., Zimmet, P., Wolf, E., Taylor, R., Ram, P., and King, H.,** Plasma uric acid and its association with diabetes mellitus and some biologic parameters in a biracial population of Fiji, *Am. J. Epidemiol.*, 127(2), 321, 1988.
20. **Ogino, N., Yamamoto, S., Hayaishi, O., and Tokuyama, T.,** Isolation of an activator for prostaglandin hydroperoxidase from bovine vesicular gland cytosol and its identification as uric acid, *Biochem. Biophys. Res. Commun.*, 87, 184, 1978.
21. **Herman, J. B.,** Hyperglycemia and uric acid, *Isr. J. Med. Sci.*, 5, 1048, 1969.
22. **Parke, A. L. and Hughes, G. R. V.,** Rheumatoid arthritis and food: a case study, *Br. Med. J.*, 282, 2027, 1981.
23. **Moment, G. B.,** Aging, arthritis and food allergies: a research opportunity revisited (editorial), *Growth*, 44, 155, 1980.

24. **Russell, A. S.,** Arthritis, inflammatory bowel disease and histocompatibility antigens, *Ann. Intern. Med.,* 86, 820, 1977.
25. **Srivastava, K. C.,** Effect of onion and ginger consumption on platelet thromboxane production in humans, *Prostaglandins Leukotrienes Essential Fatty Acids,* 35, 183, 1989.
26. **Merck Index, 10th ed.,** Merck & Co., Rahway, NJ, 1983.
27. **Koshihara, Y., Tomohiro, M., Murata, S., Lao, A., Fujimoto, Y., and Tatsuno, T.,** Selective inhibition of 5-lipoxygenase by natural compounds isolated from Chinese plants, *Artemisia rubripes nakai, F.E.B.S. Lett.,* 158, 41, 1983.
28. **Little, C. H., Stewart, A. G., and Fennessy, M. R.,** Platelet serotonin release in rheumatoid arthritis: a study in food-intolerant patients, *Lancet,* 2, 297, 1983.

5

Neurological Disorders

I. DEPRESSION

Chemical imbalances are now being recognized by some in the medical profession as a major cause of depression.[1-3] Erickson[1] has observed that chemical depression will affect from 10 to 30% of the population directly and millions more indirectly. Many of these problems go undiagnosed. Psychotherapy, cognitive therapy, shock treatment, and related methods have been used as the traditional approach to the problem of depression. They may be effective for such problems as loss of a loved one, loss of a job, or emotional disappointments and frustrations of any form, but they do not cure the individual suffering from chemical depression. In fact, such treatments may magnify the problem inasmuch as the individual's self-image may be lowered and he could experience more frustrations which could lead to suicide.

Symptoms of depression have been described by Erickson[1] as follows:

1. Emotional instability
2. Sleep disturbances
3. Loss of energy, or fatigue
4. Difficulty performing tasks
5. Loss of interest or pleasure
6. Change in appetite and weight
7. Lowered self-esteem
8. Reduced ability to concentrate
9. Thoughts of suicide
10. Physical complaints

Dr. Erickson, in his book *Depression is Curable,*[1] has concluded that a person who is moody and has four or more of the above symptoms for longer than 2 weeks has depression.

Causes of chemical imbalances which might result in depression are included in the material which follows.

Neural catecholamine exhaustion was identified by Jacobsen[4] as a cause of depression. However, the catecholamine depletion hypothesis of major depression has not been supported in more recent test results.[5,6] In fact, an opposite conclusion has been reached: in major depression, the locus ceruleus-norepinephrine system is activated.[6] Normal or increased levels of norepinephrine in cerebrospinal fluid, increased plasma norepinephrine, and increased levels of the norepinephrine metabolite 3-methoxy-4-hydroxyphenylglycol were detected in cerebrospinal fluid and urine of patients with major depression.[5] Xanthines (i.e., caffeine, theobromine, theophylline, and xanthine) promote the release of catecholamines from the adrenal medulla and peripheral adrenergic nerve endings.[7,8] Ingestion of 225 mg of the bronchodilator theophylline twice daily by a 19-year-old asthmatic girl caused her to become depressed and irritable, with frequent episodes of crying.[8] Her depression disappeared immediately after the use of theophylline was stopped. Xanthine is in such common food items as potatoes and yeast.[9] An individual who has a high chemical sensitivity could be adversely affected and become depressed by such exposure.

Primary depression has likewise been attributed to substantial hypercortisolism.[10-12] It has been suggested that the hypercortisolism is due to an abnormality at or above the level of the hypothalamus that leads to the hypersecretion of endogenous corticotropin releasing hormone.[11] This then stimulates the pituitary gland to produce the adrenocorticotropic hormone (ACTH). ACTH next triggers the production of cortisol by the adrenal gland. The cause of the abnormality has not been ascertained; certain chemicals are possibly responsible.

Homovanillic acid has been found to be elevated in severely depressed women.[10] Plasma cortisol was likewise elevated in these women.[10] Homovanillic acid is a major metabolite of dopamine, another neurotransmitter.

Lithium has been identified as a chemical which inhibits the phosphatidylinositol cycle, resulting in prophylactic action in some forms of depression.[13] The phosphatidylinositol system may become overactive due to excesses of neurotransmitters such as norepinephrine, serotonin, acetylcholine, histamine, and several peptides which activate the system.[14] Surpluses of inositol and phosphates could also accentuate this problem in patients with overactive phosphatidylinositol systems. It is possible that thyroxine also stimulates this phosphatidylinositol cycle either directly or indirectly (via β-adrenergic stimulation), since lithium has also been used in treating patients with hyperthyroidism.[14] The ability of lithium to prevent recurrent manic-depressive episodes could also be due to its ability to stabilize dopaminergic receptor sensitivity.[15]

Cerebral imbalances of serotonin, dopamine, and possibly other neurotransmitters have been cited in the etiology of depression.[16,17] Tricyclic antidepressant medications have been developed to block the reuptake of both norepinephrine and serotonin in order to keep higher amounts of these neurotransmitters circulating in the brain.[18] Some depression may be caused by excessive transmission of serotonin, and antidepressants block this transmission.[19-21]

A suggestion has been made that it is not the amount of catecholamines that is responsible for depression, but that postsynaptic membranes are unable to regulate their sensitivity to catecholamines.[22]

A deficiency in the brain chemical transmitter serotonin has been identified as a potential cause of suicide.[23,24] Asberg and co-workers[24] reported that 40% of a group of depressed patients who had attempted suicide had below normal levels of 5-hydroxyindoleacetic acid, the major breakdown product of serotonin. Francis[25] noted that reduced serotonin takeover might be a biochemical marker for suicidal and aggressive behavior. Those working in this field are attempting to determine whether lower levels of serotonin are a cause or an effect of depression.

Norepinephrine and catecholamine derivatives have been detected in many plant families, particularly in members of the Leguminosae, Musaceae, Passifloraceae, Portulaceae, Rosaceae, Rutaceae, and Solanaceae families.[26,27] The norepinephrine in bananas is formed by the β-hydroxylation of dopamine.[26] Quantities of precursors of neurotransmitters vary from food to food. The amino acid tyrosine is converted to dihydroxyphenylamine (DOPA) in the brain. The DOPA is then converted by enzyme action to dopamine and thence to norepinephrine.

The association of depression and related neurological symptoms with food ingestion was documented by McGovern et al.[28] They reported these cognitive disruptions of certain patients following food ingestion: (1) decreased concentration, memory, and mental acuity; (2) central nervous system (CNS) depression with withdrawal — depressive reactions and suicidal urges; and (3) CNS elevation — agitation and hyperkinetic reactions with violent impulses. These patients demonstrated the same symptoms when challenged with phenolic compounds. Immunological studies of the patients revealed decreased IgE with negative RAST and elevated prostaglandin $F_{2\alpha}$. Others have observed increased production of prostaglandins in association with depression and lessening of depression when antiprostaglandins were used on patients who responded poorly, or not at all, to antidepressant medications and psychotherapy.[29,30]

Light therapy has been used in an attempt to help depressed patients. An association between short daylight hours and depression prompted such investigations. The explanation for responses to light therapy is a change in the production of melatonin by the pineal gland in the brain.[31] Serotonin has been identified as the neurohormone most responsible for sleep patterns,[32] and melatonin is a by-product of serotonin metabolism.[31] Its synthesis is increased in darkness and decreased in light. The sympathetic nerves to the pineal gland regulate the circadium rhythm of melatonin synthesis by regulating the activity of the *N*-acetyltransferase, which catalyzes the formation of *N*-acetyl-5-hydroxytryptamine in the gland. Norepinephrine released from the nerve endings acts via a β-adrenergic receptor and cAMP to increase the activity of the enzyme and thus increases melatonin synthesis and secretion. Accordingly, elevated catecholamines and related cAMP increases from allergic reactions could boost the level of melatonin, leading to depression.

An antidepressant action of a calcium channel antagonist has been described in rats.[33] The mechanism responsible for this action is not fully understood, although the association of calcium with eicosanoid production, discussed in Chapter 1 and other chapters of this text, does offer a cause-effect relationship.

II. HYPERACTIVITY, LEARNING DISABILITIES, AND MEMORY

Controversy has been associated with the concept that certain foods and food additives are causative factors of hyperactivity and learning disabilities.[34-39] Estimates indicate that 4 to 10% of all U.S. school-age children have the hyperkinetic syndrome.[36] Feingold[34] first proposed that salicylates found in natural foodstuffs, synthetic food colors, and flavors cause hyperactivity and learning disabilities in susceptible children. Others have challenged that concept.[36,38] However, attention is being given to the possibility that diet may affect behavior by providing precursors for brain synthesis of neurotransmitters.[36] Salicylates as well as many synthetic food colors and flavors are compounds with aromatic (phenolic) rings. They potentiate epinephrine toxicity, presumably by interfering with the detoxification of epinephrine by methylation, since catechol *O*-methyltransferase is inhibited.[39,40] Elevated pulse and blood pressure are a sequel, with a "hyper" response resulting.

Dickerson[39] summarized world literature to 1980 on dietary factors affecting hyperactivity and learning disabilities. His conclusion was that the Feingold regimen improved the behavior of some children, but that other food allergies might also be involved. He reported that children under 5 years of age responded better to restricted diets than older children. Hypoglycemia due to a carbohydrate-rich diet was also suggested as a factor in behavioral problems.

When 76 selected overactive children were treated with an oligoantigenic diet, 62 improved and a normal behavior was achieved in 21.[37] Other symptoms, such as headaches, abdominal pain, and fits, also improved. A double-blind crossover placebo-controlled trial was conducted with 28 of the children who had improved. Foods thought to provoke symptoms were reintroduced. Symptoms returned or were exacerbated much more often when children were receiving the active material vs. the placebo. Artificial colors and preservatives were the most common provoking substances (79% reacted), but no child was sensitive to these alone. In fact, 48 foods were incriminated. Benzoic acid and tartrazine headed the list of reactive compounds, followed by soybean (73%), cow's milk (64%), and chocolate (59%). Symptoms were also noted by five patients exposed to inhalants — one with pollen, one with perfume, two with both pollen and perfume, and one with house dust. At the conclusion of this study, Egger et al.[37] noted: "This trial indicates that the suggestion that diet may contribute to behavior disorders in children must be taken seriously."

A study supporting the findings of Egger et al. described above was reported by Swain et al.,[41] whose study included 140 children and was devised for management of recurrent urticaria. They, too, found a high frequency of reactions to preservatives (including benzoates, nitrate, metabisulfite, and propionic acid), azo dyes, antioxidants, brewer's yeast, amines (tyramine and phenylethylamine), and monosodium glutamate, most children reacting to between two and five challenge compounds. The effects of these chemicals were considered to be pharmacological rather than immunological. A wise observation from the study was that "since many different foods may contain the same substances, it is easy to be misled into diagnosing 'multiple food allergy' when the common denominators have not been identified." They were able to use a standardized elimination diet with a challenge protocol to devise a suitable diet for a child within 3 months, thus lessening disruption of family life.

The influence of heavy and trace minerals on learning disabilities has been investigated.[42] Hair analysis revealed that lead and cadmium levels were significantly higher in learning-disabled children than in control children.[42]

When hyperactive children were offered a carbohydrate-rich breakfast combined with a dose of sugar, their blood sugar rose to abnormally high levels.[43] Conners[43] speculated that such a diet would cause a sudden surge in serotonin production and that hyperactive children might be more sensitive to these increases than normal children. Since glucose has been identified in the inositol triphosphate cycle, an alternative pathway is increased release of Ca^{2+} and cortisol. Conners[43] suggested that a sudden surge of serotonin might also cause imbalances in neuroagents such as dopamine and norepinephrine. Serotonin has likewise been incriminated in obsessive-compulsive disorders.[44]

Another hypothesis relating sucrose to behavioral responses comes from Langseth and Dowd.[45] They administered 5-hour glucose tolerance tests to 261 hyperactive children (ages 7 to 9) and found that 74% of the children had abnormal glucose tolerance curves. Half of the curves were suggestive of hypoglycemia. They pointed to the fact that a diet high in refined carbohydrates stimulates insulin production, which in turn stimulates an adrenergic response. This factor, coupled with hypoglycemia, is a potent stimulus for increased production of epinephrine, causing a nervous or restless action. A concomitant effect of increased production of epinephrine is the biochemical sequence leading to increased production of eicosanoids, as explained in Chapter 1.

Recent studies have increased our knowledge of the physiological role of prostaglandins in the CNS as local modulators of synaptic function.[46]

Some receptors for prostaglandins are coupled with adenylate cyclase, which is associated with the formation of cAMP.[46] Prostaglandins of the E (PGE_1) series stimulate brain tissue production and accumulation of cAMP.[45] Kanof et al.[46] noticed that platelet cAMP response to PGE_1 challenge was lower in both schizophrenic and depressive patients than in a control group. Kanof and colleagues suggested that this disorder might be due to genetically determined alterations in the coupling of the PGE receptor with adenylate cyclase or that environmental, hormonal, or plasma factors might accompany these disorders. A subsensitivity of PGE receptors in the brain would result in an impairment in the ability of locally released PGE to act as a negative feedback regulator of dopaminergic transmission. This prompted Kanof and associates to hypothesize that the resultant alterations in the functional activity of brain dopamine, norepinephrine, and possibly other neurotransmitter systems, due to impairments in local prostaglandin-mediated feedback mechanisms, might contribute to the pathophysiology of schizophrenia and major depressive disorders.

McGovern et al.[47] observed CNS effects of picogram amounts of phenolic food compounds and food extracts using sublingual provocative testing methods. Procedures were described in Section I. Abnormal CNS reactions observed included impaired concentration (41%), anger-hostility-rage syndrome (24%), mental confusion characterized by spelling and arithmetic mistakes (21%), hyperkinesis (12%), and a feeling of detachment from the body (8%). Most of the patients in this study had a history of prolonged exposure to industrial and agricultural phenolic compounds and also showed immunologic abnormalities.

Clark et al. and others[48-50] have investigated the attention deficit disorder. They have observed that this disorder is a recently reclassified pediatric neuropsychiatric disorder, characterized by inattention, impulsivity, distractibility, and sometimes hyperactivity and replaces the less formal diagnoses of hyperactivity syndrome, hyperkinetic syndrome, minimal brain dysfunction, and specific learning disability. The disorder is prevalent among preadolescent children, is reflected in poor school performance and social behavior, and has been described in the many experimental reports of impaired perceptual, cognitive, and motor functions.[51] The catecholamine (noradrenaline and dopamine) agonists methylphenidate and amphetamine are commonly used to treat the disorder, with a success rate of 80%.[52] Improvements in cognitive and social behavior are very dependent on dose levels used.[53] The conclusion of Clark et al.[48] and Raskin et al.[50] is that this disorder may be attributed to an underlying catecholamine deficiency. Evidence of abnormal levels of urinary 3-methoxy-4-hydroxyphenylethyl glycol,[54] one of the main metabolites of norepinephrine, and of reduced levels of homovanillic acid,[51] the main dopamine metabolite, in cerebrospinal fluid has been offered in support of this hypothesis. Clark et al.[48,49] also added validity to the hypothesis by using medications to block dopamine and adrenergic receptor sites. Their results implicated both dopamine and noradrenaline pathways in the regulation of attention in normal humans.[48]

The process of memory storage is still not completely understood, but many chemical modulators have been identified. Ca^{2+}-activated potassium ion (K^+) channels result in increases in K^+ permeability in surface membranes of red blood cells, neurons, alveolar macrophages, and other cells.[55-58] This response has been associated with encoding of memory. An associated interaction is that of increased influx of Ca^{2+} from an action potential which modulates the adenylate cyclase system, activating Ca^{2+} via cAMP.[56,57] Production of cAMP is, as noted in Chapter 1, increased by elevated levels of epinephrine and norepinephrine. Phenolic compounds are known to increase the release of these neurotransmitters and consequently could affect memory storage.[56] There are many other chemicals, such as metals (mercury, lead, cadmium, and chromium), which could have the same effect.[59] Lynch[60] has reported that Ca^{2+} participates in some irreversible processes by activating a class of enzymes which break down their substrate proteins rather than simply modifying them. These Ca^{2+}-activated proteinases, or calpains, are found in many types of cells. Lynch[60] suggests that the following chain of events occurs when the brain stores information in its long-term memory bank. First, high-frequency neuronal activity increases Ca^{2+} levels in the dendritic spines of a neuron and in neighboring regions, partly by allowing Ca^{2+} to rush into cells. Then, the Ca^{2+} activates the calpain, which degrades foldrin (a protein to which the enzyme attaches), disrupting the connection between the cytoskeleton of the neuron and its membrane. This in turn causes a reorganization of structure and surface chemistry at the synapse, creating new connections and uncovering additional glutamate receptors observed in experiments on long-term potentiation. Finally, the Ca^{2+} is rapidly pumped out of the dendrite, but the above effects remain, leaving the neuron more sensitive to future signals. Lynch conceded from his findings that different forms of memory have different substrates.

Impaired memory has been associated with exposure to inorganic mercury and formaldehyde.[59,61] The author of this book has learned that potassium chloride at the correct neutralizing dose will terminate a reaction to both of these chemicals. An association with potassium channels is in need of exploration. Potassium wasting and its subsequent depletion also occur as a consequence of magnesium depletion.[62] Sulfates enter the picture inasmuch as the sulfates of magnesium, sodium, and related compounds significantly increase urinary excretion of potassium.[62] This is due to the fact that in the distal nephron inorganic sulfate is nonreabsorbable, thereby obligating the excretion of an equivalent amount of cations such as sodium, potassium, or hydrogen ions.[62] Magnesium oxide is therefore recommended in preference to magnesium sulfate in cases of potassium deficits.

Deficits in central cholinergic neurotransmission result in memory impairment and reduced cognitive functioning of Alzheimer's disease.[63]

Neurochemical deficits in Alzheimer's disease include choline acetyltransferase and acetylcholinesterase, which are involved in the synthesis and catabolism, respectively, of acetylcholine.[63] Phosphate esters are considered as irreversibly acting acetylcholinesterase inhibitors due to their high affinity toward the enzyme.[64] Phosphorylation of the enzyme makes it ineffectual in the catabolism of acetylcholine; thus, increasing concentrations of acetylcholine with consequent parasympathetic excitation and severe symptoms of acetylcholine toxicosis could develop.[64] The parasympathetic activity of phosphoacetylcholine is comparable to that of acetylcholine.[64] Lecithin (phosphatidylcholine) is a possible food source with potential, as an antiacetylcholinesterase. Soybean and egg yolk are two concentrated sources of lecithin.[65]

III. CONVULSIONS AND SEIZURES

Seizures due to electroshock or convulsants are associated with an accumulation of free fatty acids (FFAs) in the brain, particularly arachidonic and stearic acid (18:0).[66] The accumulation of arachidonic acid is followed by the formation of its cyclooxygenase and lipoxygenase products.[67] Activation of phospholipase C is a prominent cause of the accumulation of FFAs in seizures, although a significant contribution from phospholipase A_2 is also a possibility.[66] Calcium is likely an activator of phospholipases, since electrical stimulation and induced seizure activity reduce extracellular Ca^{2+}, which suggests that agonist-activated calcium channels are opened, causing a rise in intracellular Ca^{2+}.[66] This is further supported from *in vitro* studies showing that the toxic effects of glutamate and other *N*-methylaspartate agonists are mediated by Ca^{2+}.[68,69]

Associations have been detected between different forms of seizures and such neurotransmitters as acetylcholine, noradrenaline, and serotonin.[66,70] Serotonin-stimulated phosphoinositide hydrolysis leads to activation of phospholipase C.[71] The result is formation of inositol-1,4,5-triphosphate, which acts as a second messenger in releasing calcium from nonmitochondrial intracellular stores, thereby elevating cytosolic calcium (Figure 5.1). Diacylglycerol is a second product formed by hydrolysis of phosphoinositides with an associated increase in arachidonic acid released from glycerol.[71] This combination of increased intracellular calcium and available arachidonic acid leads to increased formation of prostaglandins and leukotrienes via the cyclooxygenase and lipoxygenase pathways, respectively, as explained above. Such an effect is the probable cause of the seizures.

Use of methylxanthines such as theophylline for the pharmacological control of asthma has proven to be hazardous for some patients.[72-74] Convulsions, seizures, and death have been observed due to toxic effects from treatment with theophylline.[73] These effects could arise from the inhibitory effect of theophylline on phosphodiesterase activity and an effect on intracellular Ca^{2+} movement, as explained in Chapter 1. The end effect is

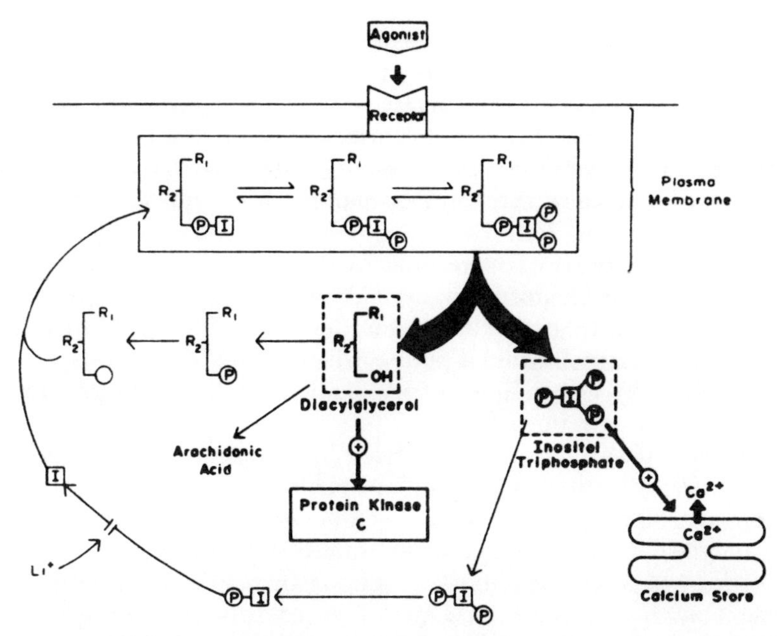

Figure 5.1 Schematic illustration of the multifunctional phospholipase C/ phosphoinositide hydrolysis signal cascade. Phosphatidyl inisitol (PI) is sequentially phosphorylated to phosphatidylinisitol-4-phosphate (PIP) and then to phosphatidylinositol-4,5-biphosphate (PIP2) by specific kinases. Agonist interaction with surface membrane receptors leads to the activation of phospholipase C, which catalyzes the hydrolysis of PIP2 to form two second messengers, inisitol triphosphate (IP3) and diacylglycerol (DAG). IP3 releases calcium from nonmitochondrial intracellular stores and thereby elevates cytosolic calcium. An important consequence of a rise in DAG levels is the release of arachidonic acid with subsequent synthesis of eicosanoids. (From Conn, P. J. and Sanders-Bush, E., *Psychopharmacology*, 92, 267, 1987. With permission.)

increased free arachidonic acid and eicosanoids, which are associated with convulsions, as outlined at the beginning of this section. Caffeine has the same effect, but is considerably less potent than theophylline.[74]

Oral intake of hydrogen cyanide (HCN), or exposure to low concentrations of HCN gas, can result in mental confusion and stupor, cyanosis, twitching, and convulsion, with terminal coma.[75]

γ-aminobutyric acid (GABA) is recognized as an inhibitory neurotransmitter in the brain and in the retina.[76] Antagonists of GABA cause convulsions.[77] Furthermore, a deficiency of pyridoxine (vitamin B_6) is associated with signs of neural hyperexcitability and convulsions inasmuch as pyridoxal phosphate is a cofactor for the enzyme glutamic decarboxylase, which decarboxylates glutamic acid to yield GABA.[76] Megadoses of B_6 to prevent convulsions are contraindicated, however, since B_6 is also a cofactor in the conversion of linoleic acid to arachidonic acid,[78] which may increase the production of eicosanoids (see above). Anticonvulsants such as phenobarbital, diphenylhydantoin, and diazepam increase the efficacy of GABA synapses.[79] Relative overactivity of excitatory amino acids (glutamic acid, for example) on postsynaptic receptors in different brain regions may induce seizures in association with a deficiency of GABA transmission.[80]

Benzodiazepines are effective anticonvulsants.[80] This effect is associated with the inhibitory neurotransmitter GABA, since the central benzodiazepine receptor includes a binding site for GABA.[80,81] Pyrethoid insecticides, such as deltamethrin and permethrin, produce convulsions in mammals, austensibly by binding to peripheral benzodiazepine receptors in nanomolar concentrations.[80]

Magnesium has anticonvulsant effects by intervening as a natural Ca^{2+} antagonist,[82] by increasing the cellular uptake of taurine, and by acting as an enzyme cofactor of cystathionine synthetase, an enzyme necessary for the catabolism of the powerful convulsant homocysteine.[83]

Taurine is a potent, selective, and long-lasting anticonvulsant.[84] This is presumably due to its role as a modulator of neuromuscular excitation, stabilizing reactivity to various stimuli, possibly by regulating Ca^{2+} flux.[85,86] Hayes et al.[85] postulated that taurine might act indirectly by modulating cellular peroxidation which activates production of such eicosanoids as thromboxane. They supported this hypothesis by demonstrating that supplementation of human diets with taurine decreased the release of thromboxane on aggregation of blood platelets and simultaneously stabilized platelets against aggregation. For these and various other physiological roles, taurine has been identified as a conditionally essential amino acid in man and in the monkey.[87]

IV. VERTIGO

Olesen[88] described the essence of vertigo as a feeling of illusionary movement, and he emphasized that vertigo is a symptom, not a disease. The illusionary movement is either rotatory or nautic. Although vertigo may arise from lesions in the CNS, Olesen observed that vertigo is more frequently labyrinthine, i.e., caused by disturbances of the vestibular end organ in the inner ear. Calcium ions are active agents in vestibular function.[89] Cerebellar calcium-binding protein has been found in the cochlear and vestibular hair cells of the rat and was assumed to prevent excessive concentrations of intracellular calcium.[89] However, in cases of excess calcium in relation to the available supply of such a binding protein, calcium-entry blockers may be deemed necessary to prevent vertigo.[88] Flunarizine and cinnarizine are two pharmaceutical agents with calcium-entry blocking activity. These complex aromatic compounds (three benzene rings per compound) have been found to be valuable drugs in the therapeutic approach to vertigo.[88,90,91] They have also been found to have antiallergic and antivasoconstricting capacities.[89]

Sodium nitroprusside is another chemical with potential to reduce cytosolic Ca^{2+}.[92] In a study of rat aorta, Magliola and Jones[92] reported that nitroprusside reduced cytosolic Ca^{2+} by combined inhibitory effects on Ca^{2+}-ATPase. This reduction in cytosolic Ca^{2+} prevented or inhibited Ca^{2+}-activated contraction and Ca^{2+}-dependent $^{42}K^+$ and $^{36}Cl^-$ efflux.

Occupational hazards which may cause vestibular-oculomotor disturbances can be found in trades such as welding, painting, and printing and in the rubber, metal, and petroleum product industries.[93] Airborne pollutants such as welding fumes and solvents give early and detectable irregularities in vestibular-oculomotor tests.[93] Vertigo or dizziness is among these early symptoms. The cause, particularly in the case of welders, has been attributed to exposure to heavy metals (Hg, Pb, Cd) in welding fumes, which have neurotoxic effects. Dizziness and disturbed equilibrium have also been reported by histology technicians exposed to formaldehyde; this persists for days after exposure.[94] The cause of this effect was not known by those reporting the study. They speculated that formaldehyde might affect neural function by condensing nonenzymatically with neuroamines, catecholamines, and indoleamines to form tetrahydroisoquinones and tetrahydro-β-carbolines, respectively.[94] The tetrahydro-β-carbolines have been shown to cause a reduction in spontaneous locomotor activity in mice.[94]

Rhinosinusitis complicated by allergic labyrinthitis manifested by nausea, dizziness, and ear pressure of 20 years' duration was cited by Girsh[95] in a case report of a 56-year-old female. Corn, cigarette smoke, grass, and wooded areas caused flare-ups. Symptoms also increased with menses.

V. TINNITUS

The figure of an estimated 40 million Americans suffering from tinnitus was reported by Meyerhoff and Mickey in 1988.[96] Yet, the underlying etiology in the majority of these patients remains obscure or conjectural.[96] Abstinence from nicotine, caffeine, and salt has benefitted some of these patients, whereas others require biofeedback, amplification, masking, and even psychotherapy.[96]

Tinnitus and/or subjective hearing loss were reported by 43% of 134 patients with rheumatoid arthritis taking regular salicylates and by 40% of untreated healthy subjects.[97] Allergic labyrinthitis manifested by tinnitus (i.e., constant ringing in the ears) had persisted in a 53-year-old male for 13 years.[95] The disorder was first noticed after the patient moved into a new house, with a change to a hot air heating system vs. a hot water system in the previous home. Increased exposure to atmospheric mold was also implicated in causation. These highly chemically sensitive patients are likely inhaling residual pollens, fungi, and molds and chemical residues in the carpeting via recirculating air; the hot water heating system would lessen air movement.

VI. SCHIZOPHRENIA

A significant number of schizophrenic patients had increased plasma PGE_2 levels in a comparison study by Kaiya et al.[98] These patients had more guilt feelings and hallucinatory behavior on the Brief Psychiatric Rating Scale, relatively successful heterosexual relations, and a higher incidence of birth complications. These researchers discussed the possibility that the elevated plasma PGE_2 levels might be due to a primary fall in PGF_1 formation, leading to increased arachidonic acid mobilization and enhanced formation of PGE_2.[98] They further proposed that schizophrenic patients with PGE_2 abnormality might possibly be homogeneous to those with PGF_1 hyposensitivity in an antagonistic relationship.

Increased formation of prostaglandins noted in schizophrenics could logically be due to elevated concentrations of norepinephrine in specific areas of the brain and in the spinal fluid.[99] Dopamine, phenylethylamine, and acetylcholine have also received attention as biochemical activating factors in this disease.[99] In fact, increased concentrations and lateral asymmetry of amygdala dopamine have been identified in schizophrenia.[100]

REFERENCES

1. **Erickson, C.,** *Depression is Curable!,* Rainbow Press, Clackamas, OR, 1986, chap. 2.
2. **Price, L. H., Charney, D. S., Rubin, A. L., and Heninger, G. R.,** α-Adrenergic receptor function in depression, *Arch. Gen. Psychiatry,* 43, 849, 1986.
3. **DiMascio, A.,** *The Benzodiazepines,* Raven Press, New York, 1973.
4. **Jacobsen, E.,** The theoretical basis of the chemotherapy of depression, in *Proceedings of the Symposium Held at Cambridge 22-26th September, 1959,* Davies, E. B., Ed., Cambridge University Press, London, 1964, 208.
5. **Gold, P. W., Goodwin, F. K., and Chrousos, G. P.,** Clinical and biochemical manifestations of depression, *N. Engl. J. Med.,* 319(6), 348, 1988.
6. **Siever, L. J. and Davis, K. L.,** Overview: toward a dysregulation hypothesis of depression, *Am. J. Psychiatry,* 142, 1017, 1985.
7. **Rall, T. W.,** Central nervous system stimulants (the xanthines), in *The Pharmacological Basis of Therapeutics,* 6th ed., Goodman, A., Goodman, L. S., and Gilman, A., Eds., Macmillan, New York, 1980, chap. 25.
8. **Murphy, M. B., Dillon, A., and Fitzgerald, M. X.,** Theophylline and depression, *Br. Med. J.,* 281, 1322, 1980.

9. *Merck Index*, 10th ed., Merck & Co., Rahway, NJ, 1981.
10. **Mazure, C. M., Bowers, M. B., Hoffman, F., Jr., Miller, K. B., and Nelson, J. C.,** Plasma catecholamine metabolites in subtypes of major depression, *Biol. Psychiatry*, 22, 1469, 1987.
11. **Kalin, N. H., Tariot, P., Barksdale, C., Weiler, S., and Thienemann, M.,** Function of the adrenal cortex in patients with major depression, *Psychiatry Res.*, 22, 117, 1987.
12. **Gold, P. W., Loriaux, L., Roy, A., Mitchel, M. A., Calabrese, J. R., Kellner, C. H., Nieman, L. K., Post, R. M., Pickar, D., Gallucci, W., Avgerinos, P., Paul, S. K., Oldfield, E. H., Cutler, G. B., Jr., and Chrousos, G. P.,** Responses to corticotropin-releasing hormone in the hypercortisolism of depression and Cushing's disease — pathophysiologic and diagnostic implications, *N. Engl. J. Med.*, 314, 1329, 1986.
13. **Menkes, H. A., Baraban, J. M., Freed, A. N., and Snyder, S. H.,** Lithium dampens neurotransmitter response in smooth muscle: relevance to action in affective illness, *Proc. Natl. Acad. Sci. U.S.A.*, 83, 5727, 1986.
14. **Mornex, R. and Orgiazzi, J. J.,** Hyperthyroidism, in *The Thyroid Gland*, DeVisscher, M., Ed., Raven Press, New York, 1980, chap. 13.
15. **Pert, A., Rosenblatt, J. E., Sivit, C., Pert, C. B., and Bunny, W. E., Jr.,** Long-term treatment with lithium prevents the development of dopamine receptor supersensitivity, *Science*, 201, 171, 1978.
16. **Freeman, R. L., Galaburda, A. M., Cabal, R. D., and Geschwind, N.,** The neurology of depression, *Arch. Neurol.*, 42, 289, 1985.
17. **Potter, W. Z.,** Introduction: norepinephrine as an 'umbrella' neuromodulator, *Psychosomatics*, 27(Suppl.), 5, 1986.
18. **Berwish, N. J. and Amsterdam, J. D.,** An overview of investigational antidepressants, *Psychosomatics*, 30, 1, 1989.
19. **Charney, D. S. and Heniger, G. R.,** Serotonin function in panic disorders, *Arch. Gen Psychiatry*, 43, 1059, 1986.
20. **Nagayama, H., Akiyoshi, J., and Tobo, M.,** Action of chronically administered antidepressants on the serotonergic postsynapse in a model of depression, *Pharmacol. Biochem. Behav.*, 25, 805, 1986.
21. **Lloyd, K. G., Farley, I. J., Deck, J. H. N., and Horneykiewicz, O.,** Serotonin and 5-hydroxyindoleacetic acid in discrete areas of the brain stem of suicide victims and control patients, *Adv. Biochem.*, 11, 387, 1974.
22. **Sulser, F.,** Deamplification of noradrenergic signal transfer by antidepressants: a unified catecholamine-serotonin hypothesis of affective disorders, *Psychopharmacol. Bull.*, 19, 300, 1983.
23. **Mann, J. J., Stanley, M., McBride, A., and McEwen, B. S.,** Increased serotonin$_2$ and β-adrenergic receptor binding in the frontal cortices of suicide victims, *Arch. Gen. Psychiatry*, 43, 954, 1986.
24. **Asberg, M., Traskman, L., and Thoren, P.,** 5-HIAA in the cerebrospinal fluid: a biochemical suicide predictor?, *Arch. Gen. Psychiatry*, 38, 1193, 1976.
25. **Francis, A.,** Reported at a conference of leading specialists on suicidal behavior, cosponsored by the New York Academy of Sciences and the National Institute of Mental Health, *Sci. Focus*, 1(1), 1, 1986.
26. **Keller, W. J.,** The involvement of epinephrine and norepinephrine in normacromerine biosynthesis, *Lloydia*, 41, 37, 1978.

27. **Robinson, T.,** *The Organic Constituents of Higher Plants — Their Chemistry and Interrelationships,* 4th ed., Cordus Press, North Amherst, MA, 1980, chap. 15.
28. **McGovern, J. J., Gardner, R. W., and Brenneman, L. D.,** Use of phenylated food compounds in diagnosis and treatment of 100 patients with food allergy and phenol intolerance, paper presented at the 37th Annu. Congr. American College of Allergists, Washington, D.C., April 4-8, 1981. *(Available in Appendix.)*
29. **Nishino, S., et al.,** Salivary prostaglandin concentrations: possible state indicators for major depression, *Am. J. Psychiatry,* 146, 365, 1989.
30. **Lowinger, P.,** Prostaglandins and organic affective syndrome, *Am. J. Psychiatry,* 146, 1646, 1989.
31. **Lewy, A. J., Wehr, T. A., Goodwin, F. K., Newsome, D. A., and Markey, S. P.,** Light suppresses melatonin secretion in humans, *Science,* 210, 1267, 1980.
32. **George, C. F. P., Millar, T. W., Hanly, P. J. and Kryger, M. H.,** The effect of L-tryptophan on daytime sleep latency in normals: correlation with blood levels, *Sleep,* 12, 345, 1989.
33. **Czyrak, A., et al.,** Some behavioral effects of repeated administration of calcium channel antagonists, *Pharmacol. Biochem. Behav.,* 35, 557, 1990.
34. **Feingold, B. F.,** Hyperkinesis and learning disabilities linked to artificial food flavours and colors, *Am. J. Nutr.,* 75, 797, 1976.
35. **Rapp, D. J.,** Food allergy treatment for hyperkinesis, *J. Learn. Disabil.,* 12, 42, 1979.
36. National Dairy Council, Diet and human behavior, *Dairy Counc. Dig.,* 52(3), 13, 1981.
37. **Egger, J., Graham, P. J., Carter, M. M., Glumley, D., and Soothill, J. F.,** Controlled trial of oligoantigenic treatment in the hyperkinetic syndrome, *Lancet,* 1, 540, 1985.
38. **Conners, C. K., Goyette, C. H., Southwick, M. A., Lees, J. M., and Andrulonis, P. A.,** Food additives and hyperkenesis: a controlled double blind experiment, *Pediatrics,* 58, 154, 1976.
39. **Dickerson, J. W. T.,** Diet and hyperactivity, *J. Hum. Nutr.,* 4, 167, 1980.
40. **Singleton, V. L.,** Naturally occurring food toxicants: phenolic substances of plant origin common in foods, *Adv. Food Res.,* 27, 149, 1981.
41. **Swain, A., Soutter, V., Loblay, R., and Truswell, A. S.,** Salicylates, oligoantigenic diets, and behaviour, *Lancet,* 2, 41, 1985.
42. **Pihl, R. O. and Parkes, M.,** Hair element content in learning disabled children, *Science,* 198, 204, 1977.
43. **Conners, K.,** The hyperactive breakfast, paper presented at the Annu. Meet. American Psychological Assoc., New York, August 28–September 1, 1987.
44. **Zohar, J., et al.,** Serotonergic responsitivity in obsessive-compulsive disorder, *Arch. Gen. Psychiatry,* 44, 946, 1987.
45. **Langseth, L. and Dowd, J.,** Glucose tolerance and hyperkinesis, *Food Cosmet. Toxicol.,* 16, 129, 1978.
46. **Kanof, P. D., et al.,** Prostaglandin receptor sensitivity in psychiatric disorders, *Arch. Gen. Psychiatry,* 43, 987, 1986.
47. **McGovern, J. J., Gardner, R. W., and Brenneman, L. D.,** The role of plant and animal phenyls in food allergy, paper presented at the 37th Annu. Congr. American College of Allergists, Washington, D.C., April 4-8, 1981. (Included in Appendix.)

48. **Clark, C. R., Geffen, G. M., and Geffen, L. B.,** Catecholamines and attention. I. Animal and clinical studies, *Neurosci. Biobehav. Rev.*, 11, 341, 1987.
49. **Clark, C. R., Geffen, G. M., and Geffen, L. B.,** Catecholamines and attention. II. Pharmacological studies in normal humans, *Neurosci. Biobehav. Rev.*, 11, 353, 1987.
50. **Raskin, L. A., et al.,** Neurochemical correlates of attention deficit disorder, *Pediatr. Clin. North Am.*, 31, 387, 1984.
51. **Hechtman, L.,** Adolescent outcome of hyperactive children treated with stimulants in childhood: a review, *Psychopharmacol. Bull.*, 21, 178, 1985.
52. **Sprague, R. L. and Sleator, E. K.,** Methylphenidate in hyperkinetic children: difference in dose effects on learning and social behavior, *Science*, 198, 1274, 1977.
53. **Shekim, W. O., Dekirmenjian, H., and Chapel, J. L.,** Urinary catecholamine metabolites in hyperkinetic boys treated with d-amphetamine, *Am. J. Psychiatry*, 123, 1276, 1977.
54. **Shaywitz, B. A., Cohen, D. J., and Bowers, M. G.,** CSF monoamine metabolites in children with minimal brain dysfunction — evidence for alteration of brain dopamine, *J. Pediatr.*, 90, 67, 1977.
55. **Blatz, A. L. and Magleby, K. L.,** Calcium-activated potassium channels, *Trends Neurosci.*, 10, 463, 1987.
56. **Kennedy, M. B.,** Molecules underlying memory, *Nature*, 329, 15, 1987.
57. **Thompson, R. F.,** The neurobiology of learning and memory, *Science*, 233, 941, 1986.
58. **Kakuta, Y., et al.,** K^+ channels of human alveolar macrophages, *J. Allergy Clin. Immunol.*, 81, 460, 1988.
59. **Rosenman, K. D., et al.,** Sensitive indicators of inorganic mercury toxicity, *Arch. Environ. Health*, 41, 209, 1986.
60. **Lynch, G.,** What memories are made of, *The Sciences*, N.Y. Academy of Science, 25, 38, 1985.
61. **Kilburn, K. H., Warshaw, R., and Thornton, J. C.,** Formaldehyde impairs memory, equilibrium, and dexterity in histology technicians: effects which persist for days after exposure, *Arch. Environ. Health*, 42, 117, 1987.
62. **Farkas, R. A., McAllister, C. T., and Blachley, J. D.,** Effect of magnesium salt anions on potassium balance in normal and magnesium depleted rats, *J. Lab. Clin. Med.*, 110, 412, 1987.
63. **Reisberg, B.,** *Alzheimer's Disease*, Macmillan, London, 1983.
64. **Wetherell, A.,** Some effects of atropine in short-term memory, *Br. J. Clin. Pharmacol.*, 10, 627, 1980.
65. **Szasz, G.,** *Pharmaceutical Chemistry of Adrenergic and Cholinergic Drugs*, CRC Press, Boca Raton, FL, 1985, chap. 3.
66. **Siesjö, B. K., Agardh, C. D., Bengtsson, F., and Smith, M. L.,** Arachidonic acid metabolism in seizures, *Ann. N.Y. Acad. Sci.*, 559, 323, 1989.
67. **Folco, G. C., Longiave, D., and Bosisio, E.,** Relations between prostaglandin E_2, $F_{2\alpha}$ and cyclic nucleotide levels in rat brain during drug-induced convulsions, *Prostaglandins*, 13, 893, 1977.
68. **Choi, D. W.,** Ionic dependence of glutamate neurotoxicity, *J. Neurosci.*, 7(2), 369, 1987.
69. **Choi, D. W.,** Glutamate neurotoxicity in cortical cell culture is calcium dependent, *Neurosci. Lett.*, 58, 293, 1985.

70. **Coleman, M.,** The effects of 5-hydroxytryptophan in Down's syndrome and other diseases of the central nervous system, in *Serotonin in Mental Abnormalities,* Boullin, D. J., Ed., John Wiley & Sons, New York, 1978, 225.
71. **Conn, P. J. and Sanders-Bush, E.,** Central serotonin receptors: effector systems, physiological roles and regulation, *Psychopharmacology,* 92, 267, 1987.
72. **Marks, M. B.,** Theophylline: primary or tertiary drug. A brief review, *Ann. Allergy,* 59, 85, 1987.
73. **Starvic, B.,** Methylxanthines: toxicity to humans. I. Theophylline, *Food Chem. Toxic.,* 26, 541, 1988.
74. **Starvic, B.,** Methylxanthines: toxicity to humans. II. Caffeine, *Food Chem. Toxic.,* 26, 645, 1988.
75. **Montgomery, R. D.,** Cyanogens, in *Toxic Constituent of Plant Foodstuffs,* Liener, I. E., Ed., Academic Press, New York, 1980, 143.
76. **Ganong, W. F.,** *Review of Medical Physiology,* 9th ed., Lange Medical, Los Altos, CA, 1979, 194.
77. **Hayashii, T.,** *Neurophysiology and Neurochemistry of Convulsions,* Dainihon-Tosho Co., Tokyo, 1959.
78. **Hamilton, E. M. N. and Whitney, E. N.,** *Nutrition — Concepts and Controversies,* West Publishing Co., New York, 1982.
79. **Roberts, E., Wong, E., Krause, D. N., Ikeda, K., and Degener, P.,** New directions in GABA research. II. Some observations relative to GABA action and inactivation, in *GABA Neurotransmitters — Pharmacochemical, Biochemical and Pharmacological Aspects,* Larsen, P. K., Scheel-Kruger, J., and Kofod, H., Eds., Academic Press, New York, 1978.
80. **Verma, A. and Snyder, S. H.,** Peripheral type benzodiazephine receptors, *Annu. Rev. Pharmacol. Toxicol.,* 29, 307, 1989.
81. **Medrum, B. and Braestrup, C.,** GABA and the anticonvulsant action of benzodiazepines and related drugs, in *Actions and Interactions of GABA and Benzodiazepines,* Bowery, N. G., Ed., Raven Press, New York, 1984, 390.
82. **Durlach, J., Bara, M., Guiet-Bara, A., and Rinjard, P.,** Taurine and magnesium homeostasis: new data and recent advances, in *Magnesium in Cellular Processes and Medicine,* Altura, B. M., Durlach, J., and Seelig, M. S., Eds., S. Karger, New York, 1985, 219.
83. **Durlach, J., Poenaru, S., Rouhani, S., Bara, M., and Guiet-Barba, A.,** The control of central neural hyperexcitability in magnesium deficiency, in *Nutrients and Brain Function,* Essman, W. B., Ed., S. Karger, New York, 1987, 48.
84. **Martin, P.,** Natural treatment for seizures — taurine, *Let's Live,* 36, 20, 1983.
85. **Hayes, K. C., Pronczuk, A., Addesa, A. E., and Stephen, Z. F.,** Taurine modulates platelet aggregation in cats and humans, *Am. J. Clin. Nutr.,* 49, 1211, 1989.
86. **Raghu, C. N., Manikeri, S. R., and Sheth, U. K.,** Probable mode of taurine action, *Indian J. Exp. Biol.,* 20, 481, 1982.
87. **Gaull, G. E.,** Taurine as a conditionally essential nutrient in man, *J. Am. Coll. Nutr.,* 5, 121, 1986.
88. **Olesen, J.,** Calcium entry blockers in the treatment of vertigo, *Ann. N.Y. Acad. Sci.,* 522, 690, 1988.
89. **Rabie, A., Thomasset, M., and Legrand, M.,** Immunocytochemical detection of calcium-binding protein in the cochlear and vestibular hair cells of the rat, *Cell Tissue Res.,* 232, 691, 1983.

90. **Ell, J. and Gresty, J.,** The effects of the "vestibular sedative" drug flunarizine upon the vestibular and oculomotor systems, *J. Neurol. Neurosurg. Psychiatry,* 46, 716, 1983.
91. **Higgs, G. A., Mugridge, K. G., and Moncada, S.,** Arachidonic acid metabolism and calcium flux, in *Calcium Entry Blockers and Tissue Protection,* Godfraind, T., Vanhoutte, P. M., Govoni, S., and Paoletti, R., Eds., Raven Press, New York, 1985, 51.
92. **Magliola, L. and Jones, A. W.,** Sodium nitroprusside alters Ca^{2+} flux components and Ca^{2+}-dependent fluxes of K^+ and Cl^- in rat aorta, *J. Physiol.,* 421, 411, 1990.
93. **Wenngren, B. I. and Ödkvist, L. M.,** Vestibulo-oculomotor disturbances caused by occupational hazards, *Acta Oto-Laryngol. (Suppl.),* 455, 7, 1988.
94. **Kilburn, K. H., Warshaw, R., and Thornton, J. C.,** Formaldehyde impairs memory, equilibrium, and dexterity in histology technicians: effects which persist for days after exposure, *Arch. Environ. Health,* 42, 117, 1987.
95. **Girsh, L. S.,** Current neurological trends with allergic implications, *Immunol. Allergy Pract.,* 185, 19, 1981.
96. **Meyerhoff, W. L. and Mickey, B. E.,** Vascular decompression of the cochlear nerve in tinnitus sufferers, *Laryngoscope,* 98, 602, 1988.
97. **Halla, J. T. and Hardin, J. G.,** Salicylate ototoxicity in patients with rheumatoid arthritis: a controlled study, *Ann. Rheum. Dis.,* 47, 234, 1988.
98. **Kaiya, H., Uematsu, M., Ofuji, M., Nishida, A., Takeuchi, K., Nozaki, M., and Idaka, E.,** Elevated plasma prostaglandin E2 levels in schizophrenia, *J. Neural Transm.,* 77, 39, 1989.
99. **Berger, P. A.,** Biochemistry and the schizophrenias — old concepts and new hypotheses, *J. Nerv. Ment. Dis.,* 169, 90, 1981.
100. **Reynolds, G. P.,** Increased concentrations and lateral asymmetry of amygdala dopamine in schizophrenia, *Nature,* 305, 527, 1987.

6

Chronic Fatigue Syndrome

A group studying chronic fatigue syndrome (CFS) offered the following description of the syndrome as a combination of nonspecific symptoms: severe fatigue, weakness, malaise, subjective fever, sore throat, painful lymph nodes, decreased memory, confusion, depression, decreased ability to concentrate on tasks, and various other complaints but with a remarkable absence of objective physical or laboratory abnormalities.[1]

Allergy has been verified as a common causative agent of CFS.[2-4] One half to three fourths of patients with the syndrome report inhalant (seasonal), food, or drug allergies.[2] A small number of these patients also report sensitivities to perfumes, solvents, cosmetics, and numerous other chemicals to which they are exposed in our contemporary environment. These patients seem to be sensitive to "everything".[5]

Metabolic disorders due to pharmacological effects of chemicals ingested and/or inhaled should most certainly be examined in the diagnosis of this syndrome since symptoms described for this syndrome are compatible with symptoms discussed in other chapters of this book. The challenge is for the physician and patient to identify specific chemical sensitivities and then take appropriate therapeutic measures.

REFERENCES

1. **Holmes, G. P., Kaplan, J. E., Gantz, N. M., Komaroff, A. L., Schonberger, L. B., Straus, S. E., Jones, J. F., Dubois, R. E., Cunningham-Rundles, C., Pahwa, S., Tosata, G., Zegans, L. S., Purtilo, D. T., Brown, N., Schooley, R. T., and Brus, I.,** Chronic fatigue syndrome: a working definition, *Ann. Intern. Med.*, 108, 387, 1988.
2. **Straus, S. E., Dale, J. K., Wright, R., and Metcalfe, D. D.,** Allergy and the chronic fatigue syndrome, *J. Allergy Clin. Immunol.*, 81, 791, 1988.
3. **Jones, L. T., Ray, C. G., Minnich, L. L., Hicks, M. J., Kibler, R., and Lucas, D. O.,** Evidence for active Epstein-Barr virus infection in patients with persistent, unexplained illnesses: elevated anti-early antigen antibodies, *Ann. Intern. Med.*, 102, 1, 1985.

4. **Straus, S. E., Tosata, G., Armstrong, M. T., Lawley, T., Preble, O. T., Henle, W., Davey, R., Pearson, G., Epstein, J., Brus, I., and Blaese, R. M.,** Persisting illness and fatigue in adults with evidence of Epstein-Barr virus infection, *Ann. Intern. Med.,* 102, 7, 1985.
5. **Zimmerman, B. and Weber, E.,** *Candida* and the "20th-century disease", *Am. Med. Assoc. J.,* 133, 965, 1985.

7

Cardiovascular Disorders

Coca[1] proposed that an increase in heart rate (pulse) of 10 beats or more per minute was indicative of an allergic reaction. Coca's observation has been repeated with the report that challenge of the antigenic heart characteristically causes tachycardia, a brief and short-lasting increase in ventricular contraction followed by prolonged contractile failure, severe dysrhythmias, and a marked decline in coronary flow.[2] In a test of 100 patients, McGovern et al.[3] reported that 15% of their patients responded with tachycardia following sublingual challenge with phenolic compounds found naturally in foodstuffs. Patients with ongoing reactions were not excluded from the pulse test, and therefore a higher percentage of patients responding with tachycardia could have been realized if the subjects had presented with a normal baseline pulse.

Prostaglandin biosynthesis by the hearts of 21 patients undergoing catheterization and coronary angiography was measured by Serneri et al.[4] This *in vivo* study disclosed that synthesis of cardiocoronary prostaglandins I_2 and E_2 (PGI_2 and PGE_2) was not appreciable under resting conditions, but became "remarkable" following stimulation via sympathetic nerves. More specifically, sympathetic stimulation was associated with a significant increase in PGI_2 and PGE_2 levels in coronary sinus, in aorta, and in peripheral venous blood. Thromboxane B_2 (TxB_2) levels likewise increased in peripheral venous blood in all but one subject following sympathetic stimulation, but not in the coronary sinus or aorta. Sympathetic stimulation was not associated with a detectable amount of $PGF_{2\alpha}$ in any vascular bed investigated.

The direct effect of nicotine of releasing catecholamine stores in the heart may be the mechanism of its acute cardiovascular effects.[5] This was in association with the increased prostaglandin production cited above[4] as well as increased synthesis of leukotrienes.

Metabolites of arachidonic acid have been implicated in cardiac dysfunction. Leukotrienes C_4 and D_4 (LTC_4, LTD_4) and prostaglandins have potent effects in the cardiovascular system, including being potent vasoconstrictors in the coronary circulation.[6-8] Changes in heart rate accompany changes in blood pressure in response to LTC_4 and LTD_4.[9] Leukotrienes also occasionally invoke release of cyclooxygenase products (prostaglandins and thromboxanes).[6] The vasoconstrictor actions of LTC_4 and LTD_4 in the coronary circulation may be involved in myocardial ischemia and angina and may be generated locally by vascular tissue, according to Piper.[6] Noting these biological actions of leukotrienes, Piper expressed great interest in seeing results of investigations of lipoxygenase inhibitors and antagonists of leukotrienes in preventing these disease states.

Another pathological effect of eicosanoids is their release in the course of coronary thrombosis and myocardial ischemia. This may modify local coronary reactivity during ischemia.[10] Lipoxygenase products, rather than cyclooxygenase products, apparently are responsible for the ischemia-induced myocardial injury. This had been postulated following a study which showed that the drug dipyridamole, which reduces infarct size, is a specific 5-lipoxygenase inhibitor.[11]

In cardiac anaphylaxis, the heart releases vasoactive substances such as histamine, PGF_2, PGE_2, PGD_2; the metabolites of TxA_2; and, of course, the slow-reacting substances of anaphylaxis (SRS-A), i.e., LTC_4, LTD_4, and LTE_4 in large amounts.[12] The anaphylactic heart is the target of these mediators released intracardially, as well as mediators reaching the left side of the heart from the lung.[12] An interaction between histamine and metabolites of arachidonic acid has been observed. Histamine may activate phospholipase A_2, causing the release of arachidonic acid and subsequent production of eicosanoids which prolong the tachyarrythmic effects of histamine, whereas PGE_2 attenuates these effects.[12]

Plasma levels of prostaglandin and thromboxane were found to be significantly ($p \leq 0.001$) elevated in patients with transient myocardial ischemia.[13] Comparative values for 6-keto-$PGF_{1\alpha}$ were 117 pg/ml vs. 10 pg/ml for normal control subjects. The prostaglandin values remained elevated through a 7-day examination period. TxB_2 values were 105 pg/ml vs. 10 pg/ml for normal control subjects. In most patients, the prostaglandin levels exceeded the thromboxane levels, but in a subgroup suffering from cardiac arrhythmias the ratio was reversed, suggesting that this ratio change might be associated with cardiac arrhythmias in myocardial infarction.[13]

Anaphylaxis has been reported as a result of ingestion of a food dye, namely tartrazine (FD&C yellow #5).[14] Desmond and Trautlein[14] indicated in their diagnosis of one patient that esophageal spasm/angioedema was a fairly prominent component. The U.S. Food and Drug Administration currently requires that food labels identify the presence of FD&C yellow #5 in food products because of several cases of reaction to this artificial food dye, which is a phenolic compound.

Hypertension reportedly affects more than half of all Americans over 65 years of age.[15] About 95% of patients with hypertension have "essential" hypertension, which means that a specific identifiable cause is not known.[16]

Salt sensitivity has been related to hypertension in elderly patients.[18] Zemel and Sowers[18] hypothesized that an age-associated increase in intracellular sodium levels in renal arteries and in leukocytes and erythrocytes is likely to result in an increase in intracellular calcium by virtue of reduced sodium-calcium exchange and thereby increased vascular resistance. This arises as a result of motor vasoconstriction of vascular muscles.[19] Calcium channel blockers are vasodilators, which may reduce this hypertensive effect.[20] A complex interaction is likely involved in the control of the level of cellular calcium, which involves more than sodium. Functions and metabolism of cAMP, prostaglandins, and calcium are interrelated,[21,22] as discussed in Chapter 1.

At least three cation transport mechanisms appear to be relevant in hypertension: bidirectional Na^+/K^+ countertransport, a process inhibited by loop diuretics; Na^+/Ca^{2+} countertransport, in which intracellular Na^+ is exchanged for Ca^{2+}; and Na^+/Li^+ countertransport.[23] Some suggest that diets high in potassium content reduce peripheral vascular resistance directly by relaxing vascular smooth muscle, but this has not always attenuated blood pressure in all models examined.[24] Magnesium is important for maintenance of cell potassium; infusions of magnesium alone have increased muscle potassium and magnesium levels and have significantly decreased the frequency of ventricular ectopic beats.[25] Wester and Dyckner[26] found that simultaneous use of both K^+ and Mg^{2+} along with reduced intake of Na^{2+} was efficacious in treating patients with essential hypertension. They proposed that one of the possible effects of Mg^{2+} might be as a natural Ca^{2+} blocker due to competition with Ca^{2+} at either the inner or outer surface of the cell membrane or through the slow channels.[27] Altered intracellular Ca^{2+} metabolism has actually been shown in human hypertension.[28] Both Ca^{2+} deficiency and excess may predispose to hypertension.[29]

Lead affects catecholamine metabolism and the renin angiotension system and induces hypertension in animals at levels below those associated with its acute toxic effects.[30] Cadmium, lead, and zinc levels were elevated in the hair of a group of adult black female hypertensives.[31] Hypertensive patients had significantly lower Zn:Cd and Cu:Zn ratios than normotensives. Of the 20 hypertensive patients, 11 were also diabetics. Glutathione is suggested as a means of protection against cadmium toxicity, since it forms complexes with heavy metals.[32] Since glutathione is part of the molecule of LTC_4 and its derivatives, LTD_4 and LTE_4, supplementation with glutathione is contraindicated, particularly when one considers that Cd supplementation has been found to decrease the activity of the selenoenzyme glutathione peroxidase.[33] Selenium has been demonstrated to have a protective effect against Cd-induced essential hypertension since it is a cofactor in glutathione peroxidase, thus reducing the level of free glutathione available for leukotriene synthesis.[34] Another possible toxicant associated with this mechanism is potassium bromate, which is included as a leavening agent in flours used in breadmaking. This compound breaks disulfide bonds; thus, the glutathione disulfide would yield two reduced molecules of glutathione which would then be available for formation of leukotrienes resulting in vasoconstriction (hypertension). The personal experience of the author suggests that supplemental selenium would also be needed to counter this effect.

Dietary increases in polyunsaturated fatty acids have been recommended to reduce hypertension and atherosclerosis.[35] Use of linoleic acid purportedly may lower blood pressure.[35] This theoretically is due to dietary linoleic acid being converted to arachidonic acid and prostanoids of the 2-ene series. Sacks et al.[36] have studied the effect of dietary linoleic acid from safflower oil on blood pressure in normotensive persons. They have concluded that variations in dietary linoleic or oleic acid are unlikely to have major effects on blood pressure or on several membrane-dependent erythrocyte functions related to hypertension.

ω-3 Fatty acids have been touted as a dietary factor which may reduce hypertension. This is based on the premise that eicosapentanenoic acid (in fish oil) competes with arachidonic acid for cyclooxygenase and gives rise to prostaglandins and thromboxanes with three instead of two double bonds.[37] It also appears that eicosapentanenoic acid selectively blocks the phospholipases for release of membrane arachidonic acid,[37] thus altering the quantities and metabolites of the lipoxygenase and cyclooxygenase pathways. Radack and Deck[38] completed a qualitative and quantitative analysis (meta-analysis) of all available randomized controlled trials which studied the effect of ω-3 fatty acids on blood pressure response. Their conclusion was that there was little scientifically sound data available to support a significant blood pressure lowering effect of ω-3 fatty acids. Caution should be exercised in recommending fish oil and related ω-3 fatty acids for use by patients with type II (non-insulin-dependent) diabetes mellitus, since a significant impairment in insulin release has been reported due to ingestion of the acids.[39]

Aromatic amines have been implicated as pressor amines which may cause severe hypertension with subsequent heart failure.[40] A "cheese reaction" results in an increase in blood pressure due to tyramine and related vasoactive amines, causing a release of norepinephrine from the sympathetic nervous system. The amines in cheese (i.e., tyramine, tryptamine, and histamine) arise via microbial amino acid decarboxylation.[40] Absorption of some amines may take place in the mouth, since reactions have been observed within 5 minutes of eating.[40] Hypertension and diabetes mellitus are common chronic conditions that frequently coexist.[41] The group studying this association found that the prevalence of hypertension was greater in persons with insulin-dependent diabetes mellitus than in those with non-insulin-dependent diabetes mellitus, after adjusting for age.[41] Essential hypertension accounted for the majority of these cases.[41] A positive correlation exists between the presence of hypertension and retinopathy in patients with diabetes mellitus.[41]

In a review lecture, Fleckenstein[44] explained why calcium antagonists are currently in increasing use worldwide in cardiovascular therapy. These compounds exert rather pronounced inhibitory effects on myocardial contractility and also act as coronary vasodilators due to their prevention of excessive transmembrane calcium uptake from sympathetic overstimulation. The most important therapeutic function of the antagonists is perhaps the effect on vascular smooth muscle, since lowering blood calcium levels suppresses phasic contractile vascular activity and tone, whereas an abnormal rise in calcium supply produces contractile hyperactivity and spasms.

The reason for the calcium effect appears to be allied with the increased production of eicosanoids (prostaglandins, leukotrienes, etc.) in response to calcium influx into the cell. The reader may refer to Chapter 1 for a discussion of the mechanism and is reminded of the discussion of eicosanoids in cardiovascular function treated earlier in this chapter.

REFERENCES

1. **Coca, A. F.,** *The Pulse Test,* Lyle Stewart, New York, 1960.
2. **Capurro, N. and Levi, R.,** The heart as a target organ in systemic allergic reactions: comparison of cardiac anaphylaxis *in vivo* and *in vitro, Circ. Res.,* 36, 520, 1975.
3. **McGovern, J. J., Gardner, R. W., and Brenneman, L. D.,** Use of phenylated food compounds in diagnosis and treatment of 100 patients with food allergy and phenol intolerance. Paper presented at the 37th Annu. Congr. American College of Allergists, Washington, D.C., April 4-8, 1981. (*Included in Appendix.*)
4. **Serneri, G. G. N. S., Gensini, G. S., Abbate, R., Prisco, D., Rogasi, P. G., Castellani, S., Casolo, G. C., Matucci, M., Fantini, F., Di Donato, M., and Dabizzi, R. P.,** Spontaneous and cold pressor test-induced prostaglandin biosynthesis by human heart, *Am. Heart J.,* 110, 50, 1985.
5. **Robertson, D., Tseng, C.-J., and Appalsamay, M.,** Smoking and mechanisms of cardiovascular control, *Am. Heart J.,* 115, 258, 1988.
6. **Piper, P. J.,** Formation and actions of leukotrienes, *Physiol. Rev.,* 64, 747, 1984.
7. **Feuerstein, G. F. and Hallenbeck, J. M.,** Leukotrienes in health and disease, *F.A.S.E.B. J.,* 1, 186, 1987.
8. **Lefer, A. M.,** Lipoxygenase-derived eicosanoids of arachidonic acid as mediators of circulatory disease states, *ISI Atl. Sci. Pharmacol.,* 2, 109, 1988.
9. **Feuerstein, G. F., Zukowska-Grojec, Z., and Kopin, I. J.,** Cardiovascular effects of leukotriene D_4 in SHR and WKY rats, *Eur. J. Pharmacol.,* 76, 107, 1981.
10. **Laurindo, F. R. M., Finton, C. K., Ezra, D., Czaja, F. J., Feuerstein, G. Z., and Goldstein, R. E.,** Inhibition of eicosanoid-mediated coronary constriction during myocardial ischemia, *F.A.S.E.B. J.,* 2, 2479, 1988.

11. **Mullane, K. M., Salmon, J. A., and Kraemer, R.,** Leucocyte-derived metabolites of arachidonic acid in ischemia-induced myocardial injury, *Fed. Proc.*, 47, 2422, 1987.
12. **Levi, R., Burke, J. A., and Corey, E. J.,** SRS-A, leukotrienes, and immediate hypersensitivity reactions of the heart, in *Advances in Prostaglandin, Thromboxane and Leukotriene Research*, Vol. 9, Samuelsson, B. and Paoletti, R., Eds., Raven Press, New York, 1982, 215.
13. **Friedrich, T., Lichey, J., Nigam, S., Priesnitz, M., and Wegscheider, K.,** Follow-up of prostaglandin plasma levels after acute myocardial infarction, *Am. Heart J.*, 109, 218, 1985.
14. **Desmond, R. E. and Trautlein, J. J.,** Tartrazine (FD&C yellow #5) anaphylaxis: a case report, *Ann. Allergy*, 46, 81, 1981.
15. **Oldfield, A. S. M.,** Epidemiologic overview, in *Blood Pressure: Regulation and Aging*, Horan, M. J. and Hadley, E. C., Eds., Biomedical Information Corp., New York, 1986.
16. **Kaplan, N. M.,** Non-drug treatment of hypertension, *Ann. Intern. Med.*, 102, 359, 1985.
17. The 1980 Report of the Joint National Committee on Detection, Evaluation and Treatment of High Blood Pressure, NIH Publ. No. 81-1088, National Institutes of Health, U.S. Health Service, Washington, D.C., December 1980.
18. **Zemel, M. B. and Sowers, J. R.,** Salt sensitivity and systemic hypertension in the elderly, *Am. J. Cardiol.*, 61, 7H, 1988.
19. **Gambert, S. R. and Duthie, E. H.,** Effect of age on red cell membrane sodium-potassium dependent adenosine triphosphatase [Na^+-K^+ ATPase] activity in healthy men, *J. Gerontol.*, 38, 23, 1983.
20. **Sowers, J. R. and Mohanty, P. K.,** Comparison of calcium-entry blockers and diuretics in the treatment of hypertensive patients, *Circulation*, 75 (Suppl. 5), 180, 1987.
21. **Cheung, W. Y.,** Calmodulin plays a pivotal role in cellular regulation, *Science*, 207, 19, 1980.
22. **McCarron, D. A.,** Is Ca more important than Na in the pathogenesis of essential hypertension?, *Hypertension*, 7, 607, 1985.
23. **Semple, P. F. and Lever, A. F.,** Glimpses of the mechanism of hypertension, *Br. Med. J.*, 293, 901, 1986.
24. **Treasure, J. and Ploth, D.,** Role of dietary potassium in the treatment of hypertension, *Hypertension*, 5, 864, 1983.
25. **Dyckner, T. and Wester, P. O.,** Potassium/magnesium depletion in patients with cardiovascular disease, *Am. J. Med.*, 82 (Suppl. 3A), 11, 1987.
26. **Wester, P. O. and Dyckner, T.,** Magnesium and hypertension, *J. Am. Coll. Nutr.*, 6, 321, 1987.
27. **Iseri, L. T. and French, J. H.,** Magnesium: nature's physiologic calcium blocker, *Am. Heart J.*, 108, 188, 1984.
28. **Blruschi, G., Blruschi, M. E., Caroppo, M., Orlandini, G., Spaggiari, M., and Cavatorta, A.,** Cytoplasmic free [Ca^{2+}] is increased in the platelets of spontaneously hypertensive rats and essential hypertensive patients, *Clin. Sci.*, 68, 179, 1985.
29. **Resnick, L. M.,** Calcium and hypertension: the emerging connection, (editorial), *Ann. Intern. Med.*, 103, 944, 1985.

30. **Lockett, C. J. and Arbuckle, D.,** Lead, ferritin, zinc, and hypertension, *Bull. Environ. Contam. Toxicol.*, 38, 975, 1987.
31. **Medeiros, D. M. and Pellum, L. K.,** Elevation of cadmium, lead, and zinc in the hair of adult black female hypertensives, *Bull. Environ. Contam. Toxicol.*, 32, 525, 1984.
32. **Singhal, R. S., Anderson, M. E., and Meister, A.,** Glutathione, a first line of defense against cadmium toxicity, *F.A.S.E.B. J.*, 1, 220, 1987.
33. **Jamall, I. S. and Smith, J. C.,** Effects of cadmium on glutathione peroxidase, superoxide dismutase, and lipid peroxidation the rat heart: a possible mechanism of cadmium cardiotoxicity, *Toxicol. Appl. Pharmacol.*, 80, 33, 1985.
34. **Christensen, M. J., Hancock, A. L., and Ford, A. H.,** Modifying effects of supplemental selenium and sulfur on cadmium toxicity in rats, *Arch. Environ. Contam. Toxicol.*, 16, 717, 1987.
35. **Iacono, M., Judd, J. T., Marshall, M. W., Canary, J. J., Dougherty, R. M., Machkin, J. F., and Weinland, B. T.,** The role of dietary essential fatty acids in reducing blood pressure, *Prog. Lipid Res.*, 20, 349, 1981.
36. **Sacks, F. M., Stampfer, M. J., Munoz, A., McManus, K., Canessa, M., and Kass, E. H.,** Effect of linoleic and oleic acids on blood pressure, blood viscosity, and erythrocyte cation transport, *J. Am. Coll. Nutr.*, 6, 179, 1987.
37. Review: fish oil and the development of atherosclerosis, Olson, R. E., Baker, D. H., and Russel, R., Eds., *Nutr. Rev.*, 45, 90, 1987.
38. **Radack, K. and Deck, C.,** The effects of omega-3 polyunsaturated fatty acids on blood pressure: a methodologic analysis of the evidence, *J. Am. Coll. Nutr.*, 8, 376, 1989.
39. **Glauber, H., Wallace, P., Griver, K., and Brechtel, G.,** Adverse metabolic effect of omega-3 fatty acids in non-insulin-dependent diabetes mellitus, *Ann. Intern. Med.*, 108, 663, 1988.
40. **Smith, T. A.,** Amines in foods, *Food Chem.*, 6, 169, 1980-81.
41. Committee report: statement on hypertension in diabetes mellitus — final report, *Arch. Intern. Med.*, 147, 830, 1987.
42. **Dollery, C. and Brennan, P. J.,** The Medical Research Council Hypertension Trial: the smoking patient, *Am. Heart J.*, 115, 276, 1988.
43. **Materson, B. J., Reda, D., Freis, E. D., and Henderson, W. G.,** Cigarette smoking interferes with treatment of hypertension, *Arch. Intern. Med.*, 148, 2116, 1988.
44. **Fleckenstein, A.,** History and prospects in calcium antagonist, *J. Mol. Cell. Cardiol.*, 22, 241, 1990.

8

Blood Glucose Levels

The consensus of an international research team studying diabetes is that at least 60% of insulin-dependent diabetes worldwide, and perhaps over 95%, is environmentally determined and thus potentially avoidable.[1] Certain viruses and chemicals are cited as causative, and they believe that the challenge for the next decade is to track down the environmental agent(s), since this is the best strategy for preventing diabetes. Thus, this group concluded that prevention must come through reducing the prevalence of environmental risk factors rather than through altering the immune system.

Non-insulin-dependent diabetes must also be considered because of extrapancreatic factors in this disorder.[2] Hypoglycemia represents a contrast and is also subject to many environmental factors. The objective of the discussion which follows is to cite factors which reportedly affect levels of blood glucose and the biological mechanisms of such effects.

Catecholamines (epinephrine, norepinephrine, and related compounds) exert a hyperglycemic effect by activation of glycogenolysis as well as gluconeogenesis.[3,4] In addition, they inhibit insulin release by activation of adrenergic receptors[4,5] and restrain glucose uptake, presumably by counteracting the effect of insulin in peripheral tissues.[2,5,6] In support of this concept is the observation that the level of circulating catecholamines is elevated in non-insulin-dependent diabetes.[7]

Hypoglycemia elicits a counterregulating hormone effect. Epinephrine release may be rapid, so the patient could experience sweating, tremor, tachycardia, anxiety, and hunger.[8] Other symptoms associated with the action of epinephrine on the central nervous system might include dizziness, headache, clouding of vision, blunted mental acuity, confusion, abnormal behavior, and in severe cases possibly convulsions and loss of consciousness.[8]

Epinephrine-induced glycogenolysis and gluconeogenesis can be decreased by using inhibitors of arachidonic acid metabolism (i.e., indomethacin and salicylates).[3] This suggests that prostaglandins interfere with hormone secretions, receptor sites, glucose transport, or other metabolic factors.[3-5,9,10] Yatomi et al.[11] determined that the relative potency of prostaglandins in producing hyperglycemia in rats was $F_{2\alpha} > D_2 > E_1 > E_2$ after injections of the prostaglandins into the third cerebral ventricle of the rats. $PGF_{2\alpha}$-induced hyperglycemia did not occur in adrenodemedullated rats.

Adverse metabolic effects of ω-3 fatty acids on non-insulin dependent diabetes mellitus give further support to the role of prostaglandins in the control of insulin secretion by pancreatic β-cells.[11] Eicosapentaenoic acid has been acclaimed as a dietary factor valuable in the prevention of atherosclerosis.[12] This is an ω-3 fatty acid derived from marine animal oils which competes with arachidonic acid for cyclooxygenase and gives rise to prostaglandins and thromboxanes with three instead of two double bonds.[12] After 1 month of a diet supplemented with fish oils, the peak insulin levels stimulated by meals or intravenous glucagon fell by 30% and 39%, respectively, in six men with type II (non-insulin-dependent) diabetes.[12] Those conducting the study urged caution when recommending ω-3 fatty acids for type II diabetic persons.[12] One of the questions raised by this study is whether a change in structure of prostaglandin, leukotrienes, or thromboxanes due to ω-3 fatty acids has a detrimental effect on β-cell function or if the "more common" cyclooxygenase products have a specific stimulatory effect.

Cyclic adenosine monophosphate (cAMP) affects glucoregulation by various mechanisms.[13] Some suggest that glucoregulation is mediated by regulation of prostaglandin synthesis at nerve endings.[10] A different mechanism suggested is that prostaglandins have receptors on liver plasma membranes which appear to be coupled to adenylate cyclase by the prostaglandins, yielding cAMP.[13] Support for the role of prostaglandins has come from the observation that salicylates have had beneficial effects on carbohydrate intolerance in diabetics.[3] Glucose itself reportedly increases the cAMP content of islets, but apparently not by direct stimulation of adenylate cyclase. The effect is mediated by calcium since there is evidence of increased intracellular Ca^{2+} concentrations in response to glucose stimulation of adenylate cyclase.[14] Release of Ca^{2+} from the endoplasmic reticulum is induced by the second messengers inositol triphosphate and arachidonic acid, as well as the guanine nucleotide GTP.[14] The evidence is that cAMP stimulates the release of insulin by increasing the concentration of Ca^{2+} in the cytosol of the β-cells.[14] The effects of cAMP are greatest at high glucose concentrations and weakest at low glucose concentrations.[14] The concentration of cAMP in islet cells is also controlled by the activity of phosphodiesterase, which catalyzes the breakdown of cAMP to AMP.[15] Inhibition of that enzyme by methylxanthines (i.e., theophylline, caffeine, and theobromine) results in substantial increases in cAMP and insulin.[15]

Large-conductance Ca^{2+}-activated K^+ channels are involved in the regulation of secretion of insulin.[16] Glucose may increase insulin secretion by decreasing the Ca^{2+}-activated K^+ current, leading to a prolongation of the plateau phase of action potentials in pancreatic β-cells. An increased duration of depolarization would increase $[Ca^{2+}]_i$ and increase secretion of insulin.[16]

Durlach and co-workers[18,19] have thoroughly researched magnesium and taurine effects on physiological processes. One of their findings was that both of these chemicals potentiated the actions of insulin, but neither enhanced insulin release from the pancreas. Taurine stimulated glycogenesis, glycolysis, and oxygen utilization in soft tissues, in addition to increasing glucose clearance from serum in the presence of insulin.

Acetylcholine stimulates insulin release in the presence of glucose.[20] Ca^{2+} is necessary for acetylcholine-stimulated insulin release, indicating that increased Ca^{2+} uptake is a necessary part of activation.[20]

An inhibition of glucose-induced insulin secretion by serotonin is another factor which has received attention.[21] An associated effect between serotonin and PGE_1 in the inhibition of insulin secretion in dogs was demonstrated by Robertson and Guest[22] when they found that methysergide, a serotonin antagonist, could reverse the inhibitory effect of PGE_1 in the dog.

Brain tryptophan levels were decreased about 40% in diabetic rats compared to controls when the rats were made diabetic by streptozotocin.[23] However, the brain serotonin levels of the diabetic rats remained unchanged. Fung et al.[23] encouraged further studies of tryptophan metabolism in diabetic patients. In contrast, an investigation was conducted to determine if insulin-induced hypoglycemia would affect the levels of serotonin and its major breakdown product, 5-hydroxyindoleacetic acid, in the hypothalamus and thalamus.[24] Levels remained unchanged, except with more severe hypoglycemia the level had risen by 35%. When these scientists discussed the increase in serotonin turnover related to insulin directly, they suggested that alternative factors should be considered, such as low blood glucose levels, secondary effects of hypoglycemia such as adrenaline and growth hormone release, or seizure activity.

Wurtman[25] has proposed that the conversion of tryptophan to serotonin is influenced by the proportion of carbohydrate in the diet; the synthesis of serotonin in turn affects the proportion of carbohydrate an individual subsequently chooses to eat.

Indomethacin antagonism of ethanol-induced hypoglycemia confirms the hypothesis that the effect of ethanol on blood glucose is mediated via prostaglandin synthesis.[26]

The physiological effects of phloridzin (phlorizin, phlorhizin, phlorrhizin, or phloretin) were reported approximately 100 years ago.[27] This molecule is one of prime interest in the study of food intolerances and associated blood glucose disorders. Glucosuria was one of the first effects observed due to lowering of the renal threshold.[2,27] Mild hypoglycemia and hypoinsulinemia are also associated pharmacological effects of phloridzin.[2] Inhibition of the transport of glucose across the intestinal mucosa by phloridzin is apparently responsible for part of this effect,[2,28] and glucagon may play a role in mild hypoglycemia induced by phloridzin.[2] Phloridzin can be found in the bark of apple, pear, plum, and cherry trees (botanical family Rosaceae). Other foods likewise appear to contain this chemical or a counterpart chemical, since it has surfaced in beet sugar, gelatin, grape, lettuce, orange, parsnip, soybean, sweet potato, yam, and yeast.

Chronic nicotine administration is accompanied by significant decreases in circulating insulin levels, but in one short study glucose levels remained unchanged.[29] Those conducting the study surmised that the chronic nicotine exposure resulted in increased fat, protein, and glycogen utilization.[29] This was due initially to increased circulating levels of catecholamines. Smoking has been associated with complications among diabetic females, particularly among pregnant females.[30] Suppressed pancreatic function has been attributed to subacute exposure to cadmium.[31] This was evidenced by lowered serum immunoreactive insulin in the presence of hyperglycemia. Magnifying the problem of reduced insulin production due to cadmium is increased production of cAMP, which activates a series of reactions resulting in the breakdown of glycogen into glucose, giving rise to hyperglycemia.[31] Gluconeogenesis from noncarbohydrate precursors has also been implicated as part of the toxic effect. Glycosuria is a consequence of the hyperglycemia.

Chromium in a biologically active form enhances the action of insulin, and therefore less insulin is required.[32] Supplemental trivalent chromium improved glucose tolerance of the elderly, adult-onset diabetics, marginally hyperglycemic subjects, and hypoglycemics.[32] Stress appears to alter chromium metabolism in animals and man.[32] Chromium seems to be mobilized from body stores into blood in response to stress and is excreted via the urine. Thus, those with blood glucose disorders may need supplemental chromium.[32]

In a review article,[33] the author(s) raised this question: "Could it be that the effects of chromium are the result of nonspecific pharmacologic effects on certain effector systems related to glucose tolerance?" For support of this implication, the author(s) noted that chromium ions have such nonspecific effects on the enzyme phosphoglucomutase.

Corticosteroids are involved in the regulation of blood glucose. Hypoglycemia activates the pituitary-adrenal axis in a normal subject, with a rise in plasma cortisol.[34] Chronic stress results in the selective desensitization of α-1 receptors involved in the potentiation of β-noradrenergic receptors to produce cAMP, therefore causing a partial reduction in the overall cAMP response. The physiological factor responsible for this effect of stress appears to be an elevated release of corticosterone.[35] Glucocorticoids have been shown to reverse the desensitization of α-1 receptors.[36]

Increased sodium-lithium countertransport activity in red cells of patients with insulin-dependent diabetes and nephropathy has been reported.[37] Some have suggested that the increase in activity of sodium-lithium countertransport in red cells may reflect an increase in activity of the brush-border sodium-hydrogen exchanger in the renal tubule.[37] According to Mangili et al.,[37] this is consistent with the finding of enhanced sodium retention in patients with diabetes. Subjects with essential hypertension also have abnormal sodium-lithium countertransport activity in red cells.[39] Perhaps lithium therapy, using proper amounts, would be of value in each case. Future research to establish these relationships would be valuable.

Certain food preservatives may cause diabetes. Nitrates and nitrites added to meats in a smoke-curing process were found to be diabetogenic.[40] *N*-nitroso compounds are supposedly the diabetogenic factor.[40] Helgason et al.[40] have hypothesized that nitrosamines act synergistically with other mutagenic agents, such as the polycyclic hydrocarbons, and these have been found in trace amounts in the meat. This problem was identified in sons of Icelandic parents who had consumed smoked, cured mutton containing *N*-nitroso compounds around the time of conception.[40]

Lillioja et al.[41] have reported that glucose itself can have adverse effects on the pancreas, citing the fact that experimental hyperglycemia can worsen abnormalities of pancreatic function in humans. They further comment that it seems equally possible that the β-cells of all persons are susceptible to "glucose toxicity" and that only those with some intrinsic abnormality of β-cells become unresponsive and thus diabetic.

A reduced product of glucose is sorbitol. Quercetin and several other flavonoids are powerful inhibitors of aldose reductases,[42] which yield sorbitol and related alcohol derivatives via the reductase reaction. Sorbitol is a compound found in foodstuffs such as cherries, plums, pears, apples, and blackstrap molasses. It is also used as a sweetener in food additives. The author determined that he was reacting to sorbitol and that the neutralizing dose of sorbitol would neutralize a reaction to glucose, but not vice versa (see Chapters 14 and 15). Does sorbitol have a toxic effect in susceptible individuals? This will have to be determined in challenge studies.

A chemical which may have an effect on the availability of the glucose found in starch is phytic acid (inositol hexaphosphate). This is due to its inhibitory effect on α-amylase.[43] Phytic acid is found in cereal grains and is often obtained from steep liquor phytin.[44] Another possible detrimental effect of phytic acid is that it chelates calcium, resulting in hypocalcemia. This could cause an imbalance in the calcium-activated potassium channels associated with insulin secretion, as discussed earlier in the chapter. Saltiel and Cuatrecasas[45] proposed that insulin stimulates the generation of a glycan containing inositol phosphate, glucosamine, and other carbohydrates. Thus, another role of phytic acid is possible in the mechanism of insulin action.

Zusman et al.[46] have postulated that the ability of chlorpropamide (a phenolic compound) to enhance glucose-stimulated insulin release in patients with diabetes mellitus is due to the inhibition of pancreatic prostaglandin synthesis, just as the ability of aspirin to increase insulin release is secondary to the inhibition of prostaglandin synthesis.[47]

REFERENCES

1. **LaPorte, R. E., Dorman, J. S., Orchard, T. J., Becker, D. J., Drash, A. L., Tajima, N., Ekoe, J-M., Tuomilehto, J., Rewers, M., Zimmet, P., Karp, M., Mohan, V., and Lee, H. K.,** Preventing insulin dependent diabetes: the environmental challenge, *Br. Med. J.*, 295, 479, 1987.
2. **Vranic, M., Glauthier, C., Bilinski, D., Wasserman, D., Tayeb, K. E., Ketenyi, G., Jr., and Lickley, H. L. A.,** Catecholamine responses and their interactions with other glucoregulatory hormones, *Am. J. Physiol.*, 247, E145, 1984.
3. **Miller, J. D., Ganguli, S., Artal, R., and Sperling, M. A.,** Indomethacin and salicylate decrease epinephrine-induced glycogenolysis, *Metabolism*, 34, 148, 1985.
4. **Turk, J., Hughes, J. H., Easom, R. A., Wolf, B. A., Scharp, D. W., Lacy, P. E., and McDaniel, M. L.,** Arachidonic acid metabolism and insulin secretion by isolated human pancreatic islets, *Diabetes*, 37, 992, 1988.
5. **Baron, A. D., Wallace, P., and Olefsky, J. M.,** *In vivo* regulation of non-insulin mediated and insulin-mediated glucose uptake by epinephrine, *J. Clin. Endocrinol. Metab.*, 64, 889, 1987.
6. **Cryer, P. E. and Gerich, J. E.,** Relevance of glucose counterregulatory systems to patients diabetes: critical roles of glucagon and epinephrine, *Diabetes Care*, 6, 95, 1983.
7. **Feldberg, W., Pyke, D. A., and Stubbs, W. A.,** On the origin of non-insulin-dependent diabetes, *Lancet*, 1, 1263, 1985.
8. **Anderson, J. A.,** Non-immunologically-mediated food sensitivity, *Nutr. Rev.*, 42, 109, 1984.
9. **Olefsky, J. M.,** Interaction between insulin receptors and glucose transport: effect of prostaglandin E_2, *Biochem. Biophys. Res. Commun.*, 75, 271, 1977.
10. **Rosen, P. and Hohl, C.,** Prostaglandins and diabetes, *Ann. Clin. Res.*, 16, 300, 1984.
11. **Yatomi, A., Iguchi, A., Yanagisawa, S., Maltsunaga, H., Niki, I., and Sakamoto, N.,** Prostaglandins affect the central nervous system to produce hyperglycemia in rats, *Endocrinology*, 121, 36, 1987.
12. **Glauber, H., Wallace, P., Griver, K., and Brechtel, G.,** Adverse metabolic effect of omega-3 fatty acids in non-insulin-dependent diabetes mellitus, *Ann. Intern. Med.*, 108, 663, 1988.
13. **Anonymous,** Fish oil and the development of atherosclerosis, *Nutr. Rev.*, 45, 90, 1987.
14. **Wolf, B. A., Colca, J. R., Turk, J., Florholmen, J., and McDaniel, M. L.,** Regulation of Ca^{2+} homeostasis by islet endoplasmic reticulum and its role in insulin secretion, *Am. J. Physiol.*, 254 (no. 2, part 1), E121, 1988.
15. **Wollhein, C. B. and Sharp, G. W. G.,** Regulation of insulin release by calcium, *Physiol. Rev.*, 61, 914, 1981.

16. **Belcher, M., Merlino, N. S., and Ro'ane, J. T.,** Control of the metabolism and lipolytic effects of cyclic 3'-5'-adenosine monophosphate in adipose tissue by insulin, methyl xanthine and nicotinic acid, *J. Biol. Chem.*, 234, 3973, 1968.
17. **Blatz, A. L. and Magleby, K. L.,** Calcium-activated potassium channels, *Trends Neurosci.*, 10, 463, 1987.
18. **Durlach, J., Bara, M., Guiet-Bara, A., and Rinjard, P.,** Taurine and magnesium homeostasis: new data and recent advances, in *Magnesium in Cellular Processes and Medicine*, Altura, B. M., Durlach, J., and Seelig, M. S., Eds., S. Karger, New York, 1987, 219.
19. **Durlach, J., Poenaru, S., Rouhani, S., Bara, M., and Guiet-Bara, A.,** The control of central neural hyperexcitability in magnesium deficiency, in *Nutrients and Brain Function*, Essman, B. M., Ed., S. Karger, New York, 1987, 48.
20. **Matthews, D. R. and Clark, A.,** Neural control of the endocrine pancreas, *Proc. Nutr. Soc.*, 46, 89, 1987.
21. **Robertson, R. P.,** Prostaglandins as modulators of pancreatic islet function, *Diabetes*, 28, 943, 1979.
22. **Robertson, R. P. and Guest, R. J.,** Reversal of methysergide of inhibition of insulin secretion by prostaglandin E in the dog, *J. Clin. Invest.*, 63, 1014, 1978.
23. **Fung, K. P., Lee, K. W., Choy, Y. M., and Lee, C. Y.,** Tryptophan and serotonin uptake of synaptosomal preparations of brains of streptozotocin-induced diabetic rats, *Nutr. Rep. Int.*, 35, 259, 1987.
24. **Gordon, A. E. and Meldrum, B. S.,** Effect of insulin on brain 5-hydroxytryptamine and 5-hydroxy-indole-acetic acid of rat, *Biochem. Pharmacol.*, 19, 3024, 1970.
25. **Wurtman, R. J.,** Ways that foods can affect the brain, *Nutr. Rev.*, 44 (Suppl.), 2, 1986.
26. **Morato, G. S., Souza, M. L. P., Pires, M. L. N., and Masus, J.,** Hypoglycemia and hypothermia induced by ethanol: antagonism by indomethacin, *Pharmacol. Biochem. Behav.*, 25, 739, 1986.
27. **Whalley, W. B.,** The toxicity of plant phenolics, in *The Pharmacology of Plant Phenolics*, Academic Press, New York, 1959, 27.
28. **Singleton, V. L. and Kratzer, F. H.,** Plant phenolics, in *Toxicants Occurring Naturally in Foods*, 2nd ed., National Academy of Sciences, Washington, D.C., 1973, chap. 15.
29. **Grunberg, N. E., Popp, K. A., Bowen, D. J., Nespor, S. M., Winders, S. E., and Eury, S. E.,** Effects of chronic nicotine administration on insulin, glucose, epinephrine, and norepinephrine, *Life Sci.*, 42, 161, 1988.
30. **Dorman, J. S. and Cruickshanks, K. J.,** Cigarette smoking, insulin-dependent diabetes mellitus and health, *Diabetes*, 71A, (Suppl. 1), 34, 1985.
31. **Singhal, R. L., Merali, Z., and Hrdina, P. D.,** Aspects of the biochemical toxicology of cadmium, *Fed. Proc.*, 35, 75, 1976.
32. **Anderson, R. A., Borel, J. S., Polansky, M. M., Bryden, N. A., Majerus, T. C., and Phylis, B.,** Chromium intake and excretion of patients receiving total parenteral nutrition: effects of supplemental chromium, *J. Trace Elements Exp. Med.*, 1, 9, 1988.
33. **Anonymous,** Is chromium essential for humans?, *Nutr. Rev.*, 46, 17, 1988.
34. **Hofeldt, F. D., Dippe, S., and Forsham, P. H.,** Diagnosis and classification of reactive hypoglycemia based on hormonal changes in response to oral and intravenous glucose administration, *Am. J. Clin. Nutr.*, 25, 1193, 1972.

35. **Stone, E. A.,** Central cyclic-AMP-linked noradrenergic receptors: new findings on properties as related to the actions of stress, *Neurosci. Biobehav. Rev.,* 11, 391, 1987.
36. **Lefkowitz, R. J. and Davies, A. O.,** Regulation of β-adrenergic receptors by steroid hormones, *Annu. Rev. Physiol.,* 46, 119, 1984.
37. **Mangili, R., Bending, J. J., Scott, G., Li, L. K., Gupta, A., and Viberti, G. C.,** Increased sodium-lithium counter-transport activity in red cells of patients with insulin-dependent diabetes and nephropathy, *N. Engl. J. Med.,* 318, 146, 1988.
38. **O'Hare, J. P., Roland, J. M., Walters, G., and Corrall, R. J. M.,** Increased sodium excretion in response to volume expansion induced by water immersion in insulin-dependent diabetes mellitus, *Clin. Sci.,* 71, 403, 1986.
39. **Canessa, M., et al.,** Increased sodium-lithium countertransport in red cells of patients with essential hypertension, *N. Engl. J. Med.,* 302, 772, 1980.
40. **Helgason, T. S., Ewen, W. B., and Stowers, J. M.,** Diabetes produced in mice by smoked/cured mutton, *Lancet,* 2, 1017, 1982.
41. **Lillioja, S., Mott, D. M., Howard, B. V., Bennett, P. H., Yki-Jarvinen, H., Freymond, D., Nyomba, B. L., Zurol, F., Swinburn, B., and Bogardus, C.,** Impaired glucose tolerance as a disorder of insulin action, *N. Engl. J. Med.,* 318, 1217, 1988.
42. **Varma, S. D.,** Inhibition of aldose reductase by flavonoids: possible attenuation of diabetic complications, in *Plant Flavonoids in Biology and Medicine: Biochemical, Pharmacological, Structure-Activity Relationships,* Cody, V., Middleton, E., Jr., and Harborne, J. R., Eds., Alan R. Liss, New York, 1986, 343.
43. **Thompson, L. U. and Yoon, J. H.,** Starch digestibility as affected by polyphenols and phytic acid, *J. Food Sci.,* 49, 1228, 1983.
44. *Merck Index,* 10th ed., Merck & Co., Rahway, NJ, 1983.
45. **Saltiel, A. and Cuatrecasas, P.,** Insulin stimulates the generation from hepatic plasma membranes of modulators derived from an inositol glycolipid, *Proc. Natl. Acad. Sci., U.S.A.,* 83, 5793, 1986.
46. **Zusman, R. M., Keiser, H. R., and Handler, J. S.,** A hypothesis for the molecular mechanism of action of chlorpropamide in the treatment of diabetes mellitus and diabetes insipidus, *Fed. Proc.,* 13, 2728, 1977.
47. **Robertson, R. P. and Metz, S. A.,** Prostaglandins, the glucoreceptor, and diabetes, *N. Engl. J. Med.,* 301, 1446, 1979.

9

Ethanol and Prostaglandins

Prostaglandins have been identified as the mediators of the central, and possibly some of the peripheral, effects of alcohol.[1] Salsolinol, a tetrahydroisoquinoline, is a phenolic compound derived from ethanol; ethanol is converted to acetaldehyde by the alcohol dehydrogenase enzyme, and the acetaldehyde then combines with dopamine to form a Schiff's base intermediate and then salsolinol (Figure 9.1).[2,3] The drug disulfiram was observed to cause unpleasant side effects within 15 minutes after subjects consumed about 15 ml of alcohol.[2] Reasons for this effect were traced to interference in two pathways resulting in increased production of salsolinol, as may be noted in Figure 9.1. The action of phenolic compounds such as salsolinol in initiating the biosynthesis of prostaglandins is discussed in Chapter 1.

Further evidence of the association between ingestion of alcohol and the formation of prostaglandins was the finding that there was decreased systemic formation of prostaglandin E (PGE) and prostacyclin and unchanged thromboxane formation in alcoholics during withdrawal, as estimated from metabolites in urine.[4] Ethanol enhances conversion of the linoleic acid metabolite dihomo-τ-linolenic acid (DGLA) to PGE_1.[5] Horrobin[5] has suggested that ethanol elevates mood by raising PGE_1 and that people often drink for the mood elevation produced by increased PGE_1 formations. He has also included with his hypothesis the concept that lithium at clinically relevant concentrations has actions which are consistent with the blockade of mobilization of DGLA, which reduces the availability of DGLA for PGE_1 formation. Transient overconsumption of ethanol would lead to depletion of DGLA, the precursor of PGE_1, and hence a transient deficit of DGLA and PGE_1, which may be one of the biochemical substrates of hangover.[5]

Involvement of 5-lipoxygenase products in ethanol-induced intestinal plasma protein loss offers evidence that the cyclooxygenase and lipoxygenase pathways are both activated by ethanol.[6]

Evidence is accumulating to support the theory that ethanol produces its intoxicating and addictive effects to a significant degree through a

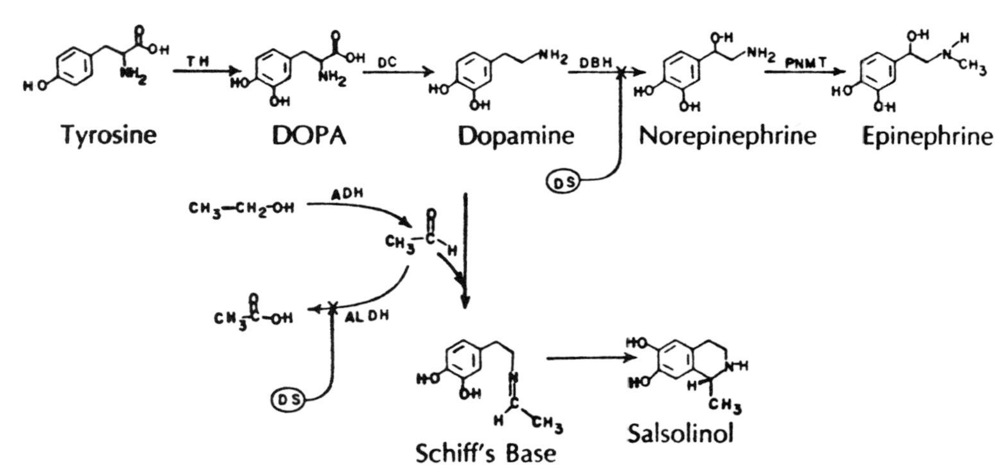

Figure 9.1 Condensation of dopamine and acetaldeyde. X = inhibition; ADH = alcohol dehydrogenase; ALDH = aldehyde dehydrogenase; D β H = dopamine β hydroxylase; DC = DOPA decarboxylase; DS = disulfiram; PNMT = phenyl-*N*-methyl transferase; TH = tyrosine hydroxylase. (From Eneanya, D. I., Bianchine, J. R., Duran, D. O., and Andresen, B. D., *Annu. Rev. Pharmacol. Toxicol.*, 21, 584, 1981. With permission.)

prostaglandin-mediated mechanism.[1] Decreased central nervous system sensitivity to ethanol produced by the use of inhibitors of prostaglandin synthesis lends credence to this theory.[1] Involvement of various neuroamines such as dopamine and serotonin in addictive behavior (specifically, alcohol craving and reward) is also proposed as an important factor.[7] The effect of ethanol on synaptosomal calcium physiology has also been studied.[8] Prostaglandins have been found to affect calcium-mediated mechanisms in nerve tissue, possibly by altering calcium availability.[8] In fact, ethanol seems to augment calcium-mediated electrophysiological events by increasing intracellular calcium concentration or effectiveness.[9] This could be associated with the prostaglandin effect, since $PGF_{2\alpha}$ induces a release of Ca^{2+} from intracellular stores.[10] Ethanol was found to suppress Ca^{2+} absorption while markedly stimulating Ca^{2+} secretion in the stomach and the distal small intestine of rats following placement of ethanol directly into the duodenum.[11] This resulted in enhanced intestinal motility. The effect of ethanol on Ca^{2+} homeostasis is perhaps at the level of the endoplasmic reticulum.[12] Davidson et al.[12] reported that Ca^{2+} channel blockers verapamil and lanthanum failed to inhibit the rise in $[Ca^{2+}]_i$ caused by ethanol. However, by preincubating synaptosomes with caffeine there was a significant decrease in the rise of $[Ca^{2+}]_i$ due to ethanol. The theory is that ethanol may diffuse through the plasma membrane and act directly on the endoplasmic reticulum.

Kuriyama et al.[13] postulated from the results of their studies that acetaldehyde might have more important pathophysiological roles than those of alcohol in the exhibition of neurotoxicity during alcohol intake. Salsolinol was cited for its modulating role on cerebral aminobutyric acid/benzodiazepine receptor complexes. The decreased capacity of such a modulating mechanism might be involved in the exhibition of alcohol withdrawal syndrome, possibly by decreasing the endogenous ligands for benzodiazepine receptors in the brain.

Both hypoglycemia and hypothermia induction by ethanol have been reported by some scientists.[14] A partial antagonism of the hypothermic effect of ethanol resulted from indomethacin (a prostaglandin synthetase inhibitor) pretreatment, and this effect was found to be ethanol dose dependent.[14] Indomethacin (5.0 mg/kg) also antagonized ethanol-induced hypoglycemia in 48-hour starved rats.[14]

Children of alcoholic mothers are at risk for a variety of birth defects involving major organ systems, as well as mental disorders.[15] This cluster of defects has been labeled the fetal alcohol syndrome (FAS).[15] Prostaglandin involvement in FAS has been convincingly demonstrated in animal model systems.[15,16] Prostaglandin synthetase inhibitors have been effective antagonists of the deleterious effects of alcohol on fetal growth.[15,16] For example, in a study of pregnant mice it was demonstrated that the teratogenic effects of ethanol on uterine-embryo tissue were attenuated by pretreatment of the dam with aspirin.[17] The aspirin reduced PGE and TxB2 levels by approximately 80 to 90%. Randall et al.[15] posed the possibility that leukotrienes might also be important in mediating the actions of alcohol on fetal growth and development. If aspirin is used, it should be used after ingestion of alcohol rather than before since aspirin significantly decreases the activity of gastric alcohol dehydrogenase.[20] The result is lessened oxidation of alcohol in the gut and, consequently, more alcohol in the circulation.[20]

REFERENCES

1. **George, F. R.,** The role of arachidonic acid metabolites in mediating ethanol self-administration and intoxication, *Ann. N.Y. Acad. Sci.*, 559, 382, 1989.
2. **Eneanya, D. I., Bianchine, J. R., Duran, D. O., and Andresen, B. D.,** The actions and metabolic fate of disulfiram, *Annu. Rev. Pharmacol. Toxicol.*, 21, 575, 1981.
3. **Smolen, T. N. and Collins, A. C.,** Behavioral effects of ethanol and salsolinol in mice selectively bred for acute sensitivity to ethanol, *Pharmacol. Biochem. Behav.*, 20, 271, 1984.
4. **Förstermann, U. and Feuerstein, T. J.,** Decreased systemic formation of prostaglandin E and prostacyclin, and unchanged thromboxane formation, in alcoholics during withdrawal as estimated from metabolites in urine, *Clin. Sci.*, 73, 277, 1987.
5. **Horrobin, D. F.,** Essential fatty acids, prostaglandins, and alcoholism: an overview, *Alcohol. Clin. Exp. Res.*, 11, 1, 1987.
6. **Beck, I. T., Boyd, A. J., and Dinda, P. K.,** Evidence for the involvement of 5-lipoxygenase products in ethanol-induced intestinal plasma protein loss, *Am. J. Physiol.*, 254, G483, 1988.
7. **Blum, K., Briggs, A. H., and Trachtenberg, M. C.,** Ethanol ingestive behavior as a function of central neurotransmission, *Experientia*, 45, 444, 1989.
8. **Anton, R. F. and Randall, C. L.,** Central nervous system prostaglandins and ethanol, *Alcohol. Clin. Exp. Res.*, 11, 10, 1987.
9. **Carlen, P. L., Gurevich, N. D., and Durand, D.,** Ethanol in low doses augments calcium-mediated mechanisms intracellularly in hippocampal neurons, *Science*, 215, 306, 1982.
10. **Altin, J. G. and Bygrave, F. L.,** Prostaglandin $F_{2\alpha}$ and the thromboxane A_2 analogue ONO-11113 stimulate Ca^{2+} fluxes and other physiological responses in rat liver, *Biochem. J.*, 249, 677, 1988.
11. **Krishnamra, N. and Limlomwongse, L.,** The *in vivo* effect of ethanol on gastrointestinal motility and gastrointestinal handling of calcium in rats, *J. Nutr. Sci. Vitaminol.*, 33, 89, 1987.
12. **Davidson, M., Wilce, P., and Shanley, B.,** Ethanol and synaptosomal calcium homeostasis, *Biochem. Pharmacol.*, 39, 1283, 1990.
13. **Kuriyama, K., Ohkuma, S., Taguchi, J., and Hashimoto, T.,** Alcohol, acetaldehyde and salsolinol-induced alterations in functions of cerebral GABA/benzodiazepine receptor complex, *Physiol. Behav.*, 40, 393, 1987.

14. **Morato, G. S., Souza, M. L. O., Pires, M. L. N., and Masur, J.,** Hypoglycemia and hypothermia induced by ethanol: antagonism by indomethacin, *Pharmacol. Biochem. Behav.*, 25, 739, 1986.
15. **Randall, C. L., Anton, R. F., and Becker, H. C.,** Alcohol, pregnancy, and prostaglandins, *Alcohol. Clin. Exp. Res.*, 11, 32, 1987.
16. **Pennington, S., Allen, Z., Runion, J., Farmer, P., and Kahmus, L.,** Prostaglandin synthesis inhibitors block alcohol-induced fetal hypoplasia, *Alcoholism Clin. Exp. Res.*, 9, 433, 1985.
17. **Anton, R. F., Becker, H. C., and Randall, C. L.,** Ethanol increases PGE and thromboxane production in mouse pregnant uterine tissue, *Life Sci.*, 46, 1145, 1990.
18. **Ross, A. D., Perlanski, E., and Grupp, L. A.,** Prostaglandin E_2 reduces voluntary ethanol consumption in the rat, *Pharmacol. Biochem. Behav.*, 36, 527, 1990.
19. **Parantainen, J.,** Prostaglandins in alcohol intolerance and hangover, *Drug Alcohol Depend.*, 11, 239, 1983.
20. **Roine, R., Gentry, T., Hernández-Muñoz, R., Baraona, E., and Lieber, C. S.,** Aspirin increases blood alcohol concentrations in humans after ingestion of ethanol, *J.A.M.A.*, 264, 2406, 1990.

10

Chemical Effects on Renal Function

In normal conditions, the ionic composition of extracellular fluid is regulated within a close range by the kidney reabsorbing or excreting sodium, potassium, other ions, and water.[1] Vasopressin is an antidiuretic hormone which regulates water excretion by increasing the water permeability of the renal collecting tubules. An increase in osmolality of extracellular fluid causes a release of vasopressin from the pituitary. The action of vasopressin appears to be mediated by activating adenylate cyclase, thereby increasing the cellular content of cyclic adenosine monophosphate (cAMP).[1] cAMP is thus mediating the effect as a "second messenger". Prostaglandins may interfere with the homeostatic mechanism of the kidney by antagonizing vasopressin-stimulated adenylate cyclase activity and water flow.[1] Prostaglandins of the E, A, and I series administered to man or experimental animals are natriuretic; i.e., they increase urine volume and urinary output of Na^+ and Cl^- ions.[2,3] Inhibition of the adenylate cyclase enzyme by prostaglandin E_2 (PGE_2) is apparently responsible for the effect inasmuch as cAMP-stimulated water flow is not affected by the prostaglandin. This physiological effect is mediated via the tubule collecting cells.[4] This is perhaps the physiological reason for the increased urinary excretion of sodium, chloride, calcium, and potassium following ingestion of caffeine.[5]

Paradoxically, vasopressin stimulates PGE_2 synthesis in the epithelial tissues of the kidney, thus simultaneously stimulating adenylate and a prostanoid which is an inhibitor of the cyclase.[1] Accordingly, PGE_2 may serve as a modulating hormone *in vivo* for the vasopressin-mediated increase in renal collecting tubule water permeability and water reabsorption.[1] However, extraneous input of chemical activators of eicosanoid production could cause an imbalance.

Leukotrienes, along with prostaglandins, have been identified and quantified in urine. Taylor et al.[6] observed that urinary excretion of leukotrienes was increased after challenge with grass antigens or in acute asthma and allergic rhinitis. They found that urinary leukotriene E_4 (LTE_4) was higher in asthmatic patients than in normal patients, although there was a substantial overlap into the normal range. The urinary LTE_4 values of the patients with rhinitis were within the normal range whether or not they had symptoms. This also provided evidence that leukotrienes are released *in vivo* in man after antigen challenge and in acute asthma. The researchers called attention to the fact that LTE_4 is a stable urinary end-product of LTC_4 and LTD_4 and, thus, LTE_4 may act as a marker of whole-body leukotriene production in man. This concept was substantiated[7] when measurable amounts of LTE_4 could be detected in the urine within 1 hour after challenge of healthy human volunteers by inhalation of amounts of LTD_4 that caused severe bronchoconstriction. LTE_4 excretion was complete after 12 hours, 50% being eliminated within 2 hours.

A fall in the excretion of PGE and prostacyclin (PGI_2) metabolites was detected in the urine of alcoholics who were abstaining.[8] No change was observed in the excretion of thromboxane A_2 (TxA_2) metabolites when compared with controls. Again, urinary levels may be used as a measure of production of eicosanoids in organisms. The association of production of prostaglandins with alcoholism is discussed in more detail in Chapter 9.

The capacity of the bladder to synthesize considerable amounts of prostanoids is influenced by pH, distension, and osmolarity.[9]

In his review paper on the biochemical mechanism of action of eicosanoids, Smith[10] summarized evidence which suggested that all prostanoids and leukotrienes operate through G protein-linked receptors in the kidney and other organs of the body. He cited two systems in which eicosanoids are involved: (1) platelet/vessel wall interaction involving PGI_2 and TxA_2, and (2) the renal water reabsorption process involving PGE_2. PGE_2 serves as an intercellular local hormone regulating a response to the circulating antidiuretic hormone (i.e., arginine vasopressin or AVP). Higher concentrations of PGE_2 (10^{-7} *M*) cause stimulation of adenylate cyclase activity in both collecting tubule and thick limb cells.[10] A feedback mechanism is involved, however, inasmuch as large increases in cAMP are known to attenuate PGE_2 production, probably by arachidonate release. The role of prostanoids in renal function was further substantiated by treatment with indomethacin (a cyclooxygenase inhibitor) or using essential fatty acid-deficient diets. The result was formation of a hyperosmotic urine since there was diminished inhibitory control by prostaglandins of the water-reabsorbing effect of the antidiuretic hormone.[11-13]

Increased appearance of PGE_3 was concurrent with a reduction of PGE_2 in the urine of subjects ingesting marine oil (n-3 fatty acids — primarily eicosapentaenoic acid).[14] Ferretti et al.[14] noted that the pharmacological properties of PGE_3 are presently largely unknown. They speculated that the alteration of the renal PGE_2/PGE_3 synthesis balance might be one of the mechanisms responsible for the well-known hypotensive effects of a marine diet. This is due to clinically relevant effects on renal excretory and uptake functions via effects on electrolyte and water regulation.

Uric acid has been identified as one of the enzyme activators in the conversion of PGG_1 to PGH_1.[15] Uric acid is apparently an activator of the prostaglandin hydroperoxidase reaction.[15] This input not only could have an impact on renal function, but also could be an important factor in the etiology of gout.

Patients with obstructive sleep apnea (OSA) often exhibit nocturnal polyuria, which disappears with nasal continuous airway pressure treatment.[16] Many patients with OSA complain of repeated nocturnal awakenings due to the need for micturition.[16] Krieger et al.[16] ascertained that patients with OSA had greater urinary flows and greater urinary sodium, chloride, and potassium excretions than did normal subjects. Nasal continuous positive airway pressure reversed these effects in patients with OSA, but had no effect on normal subjects.[16] The renal effects just described in relation to OSA may be mediated by prostaglandins since they correspond with the effects of prostaglandins discussed at the beginning of the chapter.

Exposure to xylene, toluene, and related organic solvents may cause toxic renal reactions.[17] Renal tubular acidosis together with glucosuria, bicarbonaturia, and related abnormalities known as Fanconi's syndrome may arise from this toxicosis. Willis and Elzinga[17] suggest that the high-anion-gap metabolic acidosis is likely caused by an accumulation of benzoate or hippurate.

Gossypol, a toxic organic acid contained in cottonseed, reportedly inhibits *p*-aminohippuric acid uptake in rabbit kidney slices through nonspecific nephrotoxic effects on cell metabolism and Na-K-ATPase activity.[18] Moreover, gossypol at 10^{-4} *M* significantly decreased intracellular K^+ in the slices. Hypokalemic syndrome cases were reported in clinical trials in China which those investigating suspected could have been attributed to the gossypol-induced inhibition of renal Na-K-ATPase.[19] Raw cottonseed oil was commonly used in rural communes in China in the 1950s.[20] This caused male infertility, and when gossypol was subsequently used as a contraceptive the most serious side effect was a lowering of serum potassium levels. Fatigue was a side effect, and the cause was speculated to be a prodromal symptom of a hypokalemic paralytic attack.[20] One patient developed stiffness of the neck and another developed numbness of both hands.

Caffeine and sucrose may increase urinary excretion of calcium, particularly in adolescents and college-age students.[5] Furthermore, sodium, chloride, and potassium excretions were also increased after caffeine ingestion, but not phosphorus and magnesium excretions. Combining caffeine with sucrose had an additive effect in the excretion of calcium.[5] Thus, teenagers substituting caffeine-containing soft drinks for milk may be in a negative calcium balance.[5] Increased urinary calcium excretion following caffeine ingestion may be due to chelation by 1-methyluric acid, the major metabolite of caffeine excreted in the urine.[5]

Calcium channel or entry blockers such as verapamil, nifedipine, and diltiazem appear to offer cytoprotective effects in preventing the progression of chronic renal failure,[21] in addition to their antihypertensive effects. These renal effects include an enhancement of glomerular filtration rate, renal blood flow, and electrolyte excretion.[21]

Phlorizin (phloridzin) is an aromatic compound which induces renal glucosuria by blocking tubular glucose reabsorption.[22] A fourfold increase in urine volume accompanied phlorizin-induced glycosuria in rats.[22]

REFERENCES

1. **Zusman, R. M.,** Prostaglandins and water excretion, *Annu. Rev. Med.*, 32, 359, 1981.
2. **Bolger, P. M., Eisner, G. M., and Ramwell, P. M.,** Renal actions of prostacyclin, *Nature*, 271, 467, 1978.
3. **Lote, C. J. and Haylor, J.,** Eicosanoids in renal function, *Prostaglandins Leukotrienes Essential Fatty Acids*, 36, 203, 1989.
4. **Moore, P. K.,** Prostanoids and the kidney, in *Prostanoids: Pharmacological, Physiological, and Clinical Relevance*, Moore, P. K., Ed., Cambridge University Press, New York, 1985, chap. 8.
5. **Massey, L. K. and Hollingbery, P. W.,** Acute effects of dietary caffeine and sucrose on urinary mineral excretion of healthy adolescents, *Nutr. Res.*, 8, 1005, 1988.
6. **Taylor, G. W., Black, P., Turner, N., Taylor, I., Maltby, N. H., and Fuller, R. W.,** Urinary leukotriene E_4 after antigen challenge and in acute asthma and allergic rhinitis, *Lancet*, 1, 584, 1989.
7. **Bruynzeel, P. L. B. and Verhagen, J.,** The possible role of particular leukotrienes in the allergen-induced late-phase asthmatic reaction, *Clin. Exp. Allergy*, 19 (Suppl. 1), 25, 1989.
8. **Förstermann, V. and Feuerstein, T. J.,** Decreased systemic formation of prostaglandin E and prostacyclin, and unchanged thromboxane formation, in alcoholics during withdrawal as estimated from metabolites in urine, *Clin. Sci.*, 73, 277, 1987.
9. **Mikhailidis, D. P., Jeremy, J. Y., and Dandona, P.,** Urinary bladder prostanoids. Their synthesis, function and possible role in the pathogenesis and treatment of disease, *J. Urol.*, 137, 577, 1987.
10. **Smith, W. L.,** Review article: the eicosanoids and their biochemical mechanisms of action, *Biochem. J.*, 259, 315, 1989.
11. **Anderson, R. J., Berl, T., McDonald, K. M., and Schrier, R. W.,** Evidence for an *in vivo* antagonism between vasopressin and prostaglandin in the mammalian kidney, *J. Clin. Invest.*, 56, 420, 1975.
12. **Fejes-Tóth, G., Magyar, A., and Walter, J.,** Renal response to vasopressin after inhibition of prostaglandin synthesis, *Am. J. Physiol.*, 232, F416, 1977.
13. **Hansen, H. S.,** Essential fatty acid supplemental diet increases renal excretion of prostaglandin E_2 and water in essential fatty acid deficient diet, *Lipids*, 16, 849, 1981.

14. **Ferretti, A., Flanagan, V. P., and Reeves, V. B.,** Occurrence of prostaglandin E_3 in human urine as a result of marine oil ingestion: gas chromatographic-mass spectrometric evidence, *Biochim. Biophys. Acta*, 959, 262, 1988.
15. **Ogino, N., Yamamato, S., Hayaishi, O., and Tokuyama, T.,** Isolation of an activator for prostaglandin hydroperoxidase from bovine vesicular gland cytosol and its identification as uric acid, *Biochem. Biophys. Res. Commun.*, 87, 184, 1979.
16. **Krieger, J., Imbs, J. L., Schmidt, M., and Kurtz, D.,** Renal function in patients with obstructive sleep apnea, *Arch. Intern. Med.*, 148, 1338, 1988.
17. **Willis, C. E. and Elzinga, L. W.,** Renal toxicity of xylene, *J.A.M.A.*, 261, 2258, 1989.
18. **Hong, S. K., Haspel, H. C., Sonenberg, M., and Goldinger, J. M.,** Effects of gossypol on PAH transport in the rabbit kidney slice, *Toxicol. Appl. Pharmacol.*, 71, 430, 1983.
19. **Qian, S. Z.,** Effect of gossypol on potassium and prostaglandin metabolism and mechanism of action of gossypol, in Recent Advances in Fertility Regulation, Proceedings of Symposium Organized by the Ministry of Public Health of the People's Republic of China and the World Health Organization's Special Programme of Research, Development and Research Training in Human Reproduction, Beijing, September 2-5, 1980.
20. **Liu, G. Z., Lyle, K. C., and Cao, J.,** Clinical trial of gossypol as a male contraceptive drug, *Fertil. Steril.*, 48, 459, 1987.
21. **Chan, L. and Schrier, R. W.,** Effects of calcium channel blockers on renal function, *Annu. Rev. Med.*, 41, 289, 1990.
22. **Whalley, W. B.,** The toxicity of plant phenolics, in *The Pharmacology of Plant Phenolics*, Fairbairn, J. W., Ed., Academic Press, New York, 1959, 27.

11

Sleep

The pharmacological effects of chemicals in foodstuffs or the environment on sleep offer a fruitful field for further study. For instance, dihydromethysticin found in black pepper induces a transient tranquilizing effect, deep sleep, and chaotic dreams.[1] "Disturbed" sleep was associated with increased blood pressure and pulse and changes in respiratory rate in observations by MacWilliam as early as 1923.[2] In contrast, no such changes occurred in "undisturbed" sleep.

Cow's milk allergy in children has been related to insomnia.[3] When the milk was eliminated from the diet of infants with sleep disturbances, their sleep time became similar to that of normal control children of the same age. No explanation was offered for the relation between milk allergy and sleeplessness.[3] It has been suggested that poor sleep could be related to physical discomfort, as well as to changes in metabolism of the central nervous system — through the local release of histamine, for instance. Further confusing the issue was the observation that an immediate type of allergic reaction did not seem to account for the development of poor sleep, as indicated by the absence of elevated IgE levels in the blood of most infants.

Hayaishi[4] learned that injection of prostaglandin D_2 (PGD_2) into the lateral ventricle of the brain induced sleep almost immediately in monkeys. He compared this reaction to man, for whom PGD_2 induces sleep which is indistinguishable from natural sleep. Of other prostaglandins tested, only D_3 was somewhat active, whereas $F_{2\alpha}$, D_1, 9B-D_2, and D_2 methyl ester were much less effective or inactive. An opposite effect was observed with PGE_2 since this prostaglandin has an inhibitory effect on sleep.[4] In the rat the amount of slow-wave sleep was reduced to 70% of the controls and rapid eye movement (REM) to about 40% following injection of PGE_2. A possible reason for these physiological effects on sleep may be due to effects of prostaglandins on the synthesis and release of serotonin.[5] Bhattacharya et al.[5] challenged rats with different prostaglandins to elucidate the specific action of these substances on neurotransmitters, including serotonin. PGD_2 and PGE_1 facilitated rat brain serotonergic activity; conversely, $PGF_{2\alpha}$ appeared to attenuate serotonergic activity.

Serotonin and melatonin (a metabolite of serotonin) are hormones associated with REM sleep.[6] Melatonin is produced in the pineal gland and appears to be secreted in man only during darkness.[7] A small increase in serotonin availability in the brain apparently causes an increase in sleepiness or sedation.[8,9] Lowering serotonin produces insomnia or increased arousal.[8] Perhaps this is due to a decrease in the production of melatonin. A method of increasing serotonin availability is to supplement the diet with additional tryptophan.[6,8] Leathwood found that this required 0.5 to 1.0 g or more of the amino acid.[8] Less was required when combined with a carbohydrate load.

Caffeine has been cited as a factor associated with wakefulness and reduced quality of sleep. Yet, surprisingly, when caffeine was incorporated as a pure compound (at 0.3%) in a diet fed to rats, there were significant and persistent increases in the brain concentrations of tryptophan, serotonin, and the major degradation product of the latter, namely, 5-hydroxyindoleacetic acid.[9] These conflicting observations suggest that quantities and interactions may be involved. Species differences in responses should perhaps be considered as well.

Hayaishi[4] challenged rats with different prostaglandins to elucidate the specific action of these substances on neurotransmitters, including serotonin. PGD_2 and PGE_1 facilitated rat brain serotonergic activity. $PGF_{2\alpha}$ appeared to attenuate serotonergic activity. After subsequent studies, Hayaishi[10] proposed that PGD_2 and PGE_2 are probably two of the major endogenous sleep-regulating substances — one promoting sleep and the other wakefulness — in rats, dogs, rabbits, monkeys, and probably in humans as well. He reported preliminary evidence that the site of action of PGD_2 is in or near the preoptic area and that of PGE_2 is the hypothalamus.

Functional interactions of norepinephrine (NE), serotonin (5-hydroxytryptamine [5-HT]), dopamine, and histamine have been observed to occur during physiological REM sleep.[11] After identifying these relationships, Monti[11] suggested that a future version of the reciprocal interaction model of sleep cycle control would probably include some other modulator neurons in addition to the NE and 5-HT inputs.

A painful nocturnal penile erection (PE) associated with the rapid eye movement of sleep is a sleep disorder affecting both young and old males alike.[12,13] Melatonin seems a logical chemical to ascribe as a triggering factor in this response, due to its association with the REM of sleep. However, dopamine agonists and nanogram amounts of oxytocin have induced repeated episodes of PE and yawning in male rats.[14,15] Does melatonin trigger these responses? The answer is currently not known. A possible mechanism of oxytocin induction of PE and yawning has been attributed by Argiolas and associates[15] to calcium acting as a second messenger. They compared calcium channel inhibitors (verapamil, flunarizine, nimodipine, etc.) with indomethacin and aspirin (inhibitors of prostaglandin synthesis). The calcium channel inhibitors were effective in inhibiting oxytocin induction of PE and yawning at relatively high doses, in contrast to no response from the prostaglandin inhibitors.[16] Inhibitors of the lipoxygenase enzymes (leukotriene synthesis) were not included in the study. These researchers also found that the effect of calcium channel inhibitors seemed to be central, since the response was observed not only after intraperitoneal administration of the drugs, but also after their intracerebroventricular injection.[15] The findings suggested to them that oxytocin exerts its effect on PE and yawning by increasing calcium influx in some neuronal population of the rat brain. They used both calcium channel inhibitors and inhibitors of prostaglandin synthesis inasmuch as in the uterus oxytocin receptor stimulation induces both calcium mobilization and synthesis of prostaglandins. Oxytocin may then be a therapeutic agent in reducing impotency prior to sex.

PE has also been induced in rats by the 5-HT-releasing compound fenluramine, 5-HT reuptake inhibitors, and related compounds.[16] Berendsen et al.[16] propose that 5-HT_{1c} receptors mediate the PE. Blockage of PE by 5-HT antagonists selective for the 5-HT_{1c} receptor supports the hypothesis.

Since melatonin is a catabolic metabolite of 5-HT, there may be an interplay in which either 5-HT or melatonin activates the synthesis and release of oxytocin during sleep. One of the rate-limiting steps in the synthesis of 5-HT in the brain from the amino acid tryptophan is the Ca^{2+}-dependent phosphorylation of the enzyme.[8] Is this possibly another role for Ca^{2+} in the PE response? Additional studies are needed to identify interactions. Another pharmacological effect may arise from 5-hydroxyindole-3-acetic acid found in urine as the principal catabolite of 5-HT and related indoleamines.[17]

A clinical study of a 63-year-old man with a 20-year history of repeated awakenings at night due to painful erections was conducted in a sleep laboratory.[12] Most nights he awakened with them on at least four occasions, went to the toilet to micturate, lost the erection, and returned to sleep. Dreaming was often noted prior to awakening. He complained of inadequate sleep and reported being tense and irritable during the day.

Insomnia and disturbed sleep are attributed to a variety of causes, such as physiological and psychological disorders and the aging process.[18] Monroe[19] compared physiological responses of good and poor sleepers throughout the sleep period. Poor sleepers had significantly higher mean rectal temperatures, a higher mean number of phasic vasoconstrictions during both sleep and presleep periods, higher heart rates during the presleep period, and more body movement activity and higher basal skin resistance while sleeping than good sleepers. The poor sleepers are thus exhibiting symptoms characteristic of those described in other chapters of this book, particularly Chapter 7.

Sleep apnea (either central or obstructive) results in hypoxia. Two locally formed vasodilators, prostacyclin (PGI_2) and adenosine, are released from hearts subjected to different degrees of hypoxia and ischemia.[21] Adenosine is probably the more important of the two as a vasodilating agent.[21]

REFERENCES

1. **Selner, J. C. and Staudenmayer, H.,** The relationship of the environment and food to allergic and psychiatric illness, in *Psychobiological Aspects of Allergic Disorders,* Young, S. H., Rubin, J. M., and Daman, H. R., Eds., Praeger Scientific, New York, 1986, chap. 6.
2. **MacWilliam, J. A.,** Some applications of physiology to medicine. III. Blood pressure and heart action in sleep and dreams, *Br. Med. J.,* 2, 1196, 1923.
3. **Kahn, A., Rebuffat, E., Blum, D., Casimir, G., Duchateau, J., Mozin, M. J., and Jost, R.,** Difficulty in initiating and maintaining sleep associated with cow's milk allergy in infants, *Sleep,* 10, 116, 1987.
4. **Hayaishi, O.,** Prostaglandin D_2 and sleep, *Ann. N.Y. Acad. Sci.,* 559, 374, 1989.
5. **Bhattacharya, S. K., Dasgupta, G., and Sen, A. P.,** Prostaglandins modulate central serotonergic neurotransmission, *Indian J. Exp. Biol.,* 27, 393, 1989.
6. **George, C. F. P., Millar, T. W., Hanly, P. J., and Kryger, M. H.,** The effect of L-tryptophan on daytime sleep latency in normals: correlation with blood levels, *Sleep,* 12, 345, 1989.
7. **Strassman, R. J., Peake, G. T., Qualls, C. R., and Lisansky, E. J.,** A model for the study of the acute effects of melatonin in man, *J. Clin. Endocrinol. Metab.,* 65, 847, 1987.
8. **Leathwood, P. D.,** Tryptophan availability and serotonin synthesis, *Proc. Nutr. Soc.,* 46, 143, 1987.
9. **Yokogoshi, H., Tani, S., and Amano, N.,** The effects of caffeine and caffeine-containing beverages on the disposition of brain serotonin in rats, *Agric. Biol. Chem.,* 51, 3281, 1987.
10. **Hayaishi, O.,** Molecular mechanisms of sleep-wake regulation: roles of prostaglandins D_2 and E_2, *F.A.S.E.B. J.,* 5, 2575, 1991.
11. **Monti, J. M.,** Are cholinergic, noradrenergic and serotonergic neurons sufficient for understanding REM sleep control?, *Behav. Brain Sci.,* 9, 413, 1986.
12. **Matthews, B. J. and Crutchfield, M. B.,** Painful nocturnal penile erections associated with rapid eye movement sleep, *Sleep,* 10, 184, 1987.
13. **Reynolds, C. F., III, Thase, M. E., Jennings, J. R., Howell, J. R., Frank, E., Berman, S. R., Houck, P. R., and Kupfer, D. J.,** Nocturnal penile tumenescence in healthy 20- to 59-year-olds: a revisit, *Sleep,* 12, 368, 1989.
14. **Argiolas, A., Melis, M. R., Fratta, W., Mauri, A., and Gessa, G. L.,** Monosodium glutamate does not alter ACTH- or apomorphine-induced penile erection and yawning, *Pharmacol. Biochem. Behav.,* 26, 503, 1987.

15. **Argiolas, A., Melis, M. R., Stancampiano, R., and Gessa, G. L.,** Oxytocin-induced penile erection and yawning: role of calcium and prostaglandins, *Pharmacol. Biochem. Behav.*, 35, 601, 1990.
16. **Berendsen, H. H. G., Jenck, F., and Broekkamp, C. L. E.,** Involvement of 5-HT_{1c} receptors in drug-induced penile erections in rats, *Psychopharmacology,* 101, 57, 1990.
17. **Ganong, W. F.,** *Review of Medical Physiology,* 9th ed., Lange Medical Publications, Los Altos, CA, 1979, chap. 15.
18. **Anch, A. M., Browman, C. P., Mitler, M. M., and Walsh, J. K.,** *Sleep: A Scientific Perspective,* 1st ed., Prentice Hall, Englewood Cliffs, NJ, 1988, chap. 9.
19. **Monroe, L. J.,** Psychological and physiological differences between good and poor sleepers, *J. Abnorm. Psychol.,* 72, 255, 1967.
20. **Edlund, A., Fredholm, B. B., Patrignani, P., Patrona, C., Wennmalm, A., and Wennmalm, Å.,** Release of two vasodilators, adenosine and prostacyclin, from isolated rabbit hearts during controlled hypoxia, *J. Physiol.,* 340, 487, 1983.

12

Headaches

A survey reported in Britain in 1974 estimated that about 20% of the population are affected by migraine, with more women affected than men.[1] In one study, two thirds of those experiencing severe migraine were allergic to certain foods, as determined by dietary exclusion and subsequent challenge.[2] Other factors such as chain smoking and the use of corticosteroids were suspected in a group which did not respond to the food elimination diet.[2] In a second study, 93% of 88 children with severe frequent migraine recovered on oligoantigenic diets; the causative foods were identified by sequential reintroduction, and the role of the foods provoking migraine was established by a double-blind controlled trial in 40 of the children.[3] Most of the children reacted to several foods. Almost all had behavior disturbances at the time of an attack. Associated symptoms which also improved on the diets included abdominal pain, fits, asthma, and eczema. Migraine resulted from challenge with benzoic acid and tartrazine in a number of the children. The five foods causing the most responses were cow's milk, egg, chocolate, orange, and wheat. Processing of a food affected its tendency to provoke symptoms; some patients reacted to white flour but not to brown, and four reacted to bacon but not pork. Another observation was that tobacco smoke and perfume also provoked symptoms in some children. Those reporting the study suggested that IgE may not be important in the mechanism of the presumed allergy because of lack of IgE antibodies to many of the causative foods. Abdominal symptoms recurred first when patients were challenged with provoking foods, suggesting that the allergic reaction might occur in the gut with release of mediators as antigens in the circulation.

Pulse rate and blood pressure were substantially reduced by the use of an elimination diet tested on migraine patients in another study.[4] Food allergy was again found to play a major part in the provocation of migraine attacks.

Studies of the pharmacological causation of migraine have included the hypothesis that the modulation of synaptic serotonin — centrally as well as peripherally — is an important pathogenic mechanism of the disorder.[5]

Plasma serotonin rises just before the onset of a migraine headache and falls markedly during an attack.[6,7] Stress, dietary factors, hormonal changes, and hypoglycemia were cited as precipitating factors in migraine by Jones et al.,[8] who related the problem to a platelet disorder. Their observation was that platelets of migraine patients showed a gradual increase in potential release of serotonin. They associated this with increased levels of epinephrine and norepinephrine, due to certain factors, activating release of serotonin from platelets. This appeared to trigger the complex chain of vascular responses and biochemical changes that characterize the migraine attack. They determined that the platelets of migraine patients were more sensitive to serotonin-releasing agents than the platelets of normal control subjects. Tyramine administered both orally and intravenously can precipitate attacks in migraine patients, since it has a potent effect in releasing serotonin from platelets.[9,10] Human cerebral arteries show a pronounced sensitivity to serotonin.[11] L-tryptophan, derived from dietary sources, increases the rate of synthesis of brain serotonin almost immediately after the administration of this amino acid.[12] Thus, precursor availability is a factor associated with plasma levels of serotonin and related migraine.

Cheese has been identified as a food which causes severe migraine. The cause has been attributed to biogenic amines — particularly histamine, tyramine, and tryptamine.[13] These physiologically active amines are also found in common fruits and vegetables.[14] Bananas lead the list in quantities, particularly in the peel.[14] Dopamine, serotonin, and norepinephrine were amines identified in these plant foods, in addition to the aforementioned foods.

"Hot dog" headaches have been attributed to ingestion of sodium nitrite in frankfurters.[15] A patient who developed a headache after eating frankfurters was challenged with sodium nitrite (10 mg); these results were compared to those obtained using a placebo of sodium bicarbonate. Headaches were provoked 8 out of 13 times after ingestion of sodium nitrite, but never after challenge with the placebo.[15]

Exposure of migraine-susceptible patients to phenolic compounds via either oral or nasal routes should also be considered in causation of migraine. Gallic acid (a phenolic compound) potentiates the toxicity of epinephrine, a property shared with several other readily soluble phenols, presumably by interfering with the enzyme which detoxifies epinephrine by methylation.[16,17]

The physiological effects of serotonin (5-HT) in attacks of migraine may be due to release of prostaglandins, particularly PGE.[18] Support of this role comes from the observation that intravenous PGE in nonmigrainous subjects consistently resulted in a vascular headache that bore migrainous features.[5] In addition, 5-HT is one of the few compounds that release PGE into the ventricular fluid when perfused into the cerebral ventricles.[5]. An increase in 5-hydroxyindoleacetic acid, a metabolite of 5-HT, is associated with a concurrent fall in platelet 5-HT levels during food challenges.[19]

Although 5-HT has just been implicated as an activating force in the onset of headaches, increased 5-HT in the cerebral vascular system has been found effective in the reversal of the pain of migraine headaches.[20-24] Friberg et al.[20] measured regional cerebral blood flow and blood velocity in the middle cerebral arteries in migraine patients before and after intravenous infusion of sumatriptan (2 mg), a 5-HT-like receptor agonist. Observing a significantly lower blood velocity on the headache side of the brain than on the nonheadache side of the brain, they concluded that the pain of the headache was due to dilitation of the middle cerebral arteries. Infusion of sumatriptan relieved symptoms within 30 minutes without affecting regional cerebral blood flow. This suggested to them that migraine is associated with, if not caused by, a low serotoninergic drive from the perivascular nerves or a reduced number of 5-HT receptors on the smooth muscles of the arteries. An observation supporting the 5-HT hypothesis comes from the link between migraine-like headaches and use of cocaine.[25] The immediate effect of cocaine is to increase concentrations of 5-HT, but after repeated or chronic exposure the system is suppressed, resulting in dysregulation.[25,26] Perhaps 5-HT is acting as a vasoconstrictor to counter the effects of PGI_2 (prostacyclin) and/or PGE_2, which are vasodilators.[27] Cyclooxygenase inhibitors would seem a logical choice to try as an alternate form of therapy. As evidence of this are reports that an amelioration of migraine in some patients followed use of aspirin, as well as caffeine.[28] Burnstock[28] hypothesized that the aspirin would reduce the ATP-induced prostaglandin component of vasodilatation and pain, and caffeine as a purinoceptor antagonist would reduce the AMP and adenosine component.

Monosodium glutamate (MSG) has been implicated in migraine.[29,30] An estimated 30% of individuals who eat MSG-containing Chinese food develop symptoms approximately 20 minutes after ingestion.[30] In a survey, 159 individuals of 414 with "Chinese Restaurant Syndrome" reported having headaches associated with ingestion of MSG.[31] Since symptoms of reactions to MSG are almost identical to those of acetylcholinemia and since glutamate can be converted to acetylcholine, Ghadimi et al.[32,33] hypothesized a transient rise in acetylcholine in response to ingestion of MSG. This was likewise accentuated because of decreased serum cholinesterase due to MSG,[33] suggesting that the clinical responses were the result of parasympathetic stimulation. The reason some individuals react to MSG and others do not is a matter requiring further study. Other chemicals, in addition to MSG, which are suspected of causing migraine are ethanol, sodium nitrate, caffeine, phenylethylamine, tyramine, sodium metabisulfite, theobromine, and benzoic acid.[34]

At the 2nd International Congress on Headache, held in Copenhagen in June 1985, Professor T. Peters (West Germany) noted that current thinking on the pathology of migraine was related to brain hypoxia and spreading depression, which is a neural depolarization extending gradually over the cortex.[35] Cellular loss of potassium ions with a consequent increase in extracellular levels of potassium induced a pathological calcium influx.[35] Large increases in intracellular calcium concentrations overstimulated normal cell functions, inducing neurotransmitter release, phospholipase activation, and mitochondrial calcium uptake, which in turn resulted in vasospasm, neuronal dysfunction, and finally brain-cell death.[35] Flunarizine, a calcium entry blocker, was reported at that conference to be a valuable new drug in the prophylaxis of migraine. Side effects from the drug occasionally included drowsiness, lethargy, and depression. Verapamil, another calcium entry blocker, is also efficacious in migraine prophylaxis.[36]

An association between hypoglycemia and migraine has been reported.[37] Hypoglycemia reportedly affects the tone of cranial blood vessels, and headache often accompanies insulin shock.[38] Headache accompanying a fast or hunger is a common complaint. Antihypoglycemic diets have shown therapeutic value in averting migraine when patients have transient hypoglycemia.[37]

Alcoholic beverages have been associated with migraine.[39,40] Some residual chemicals soluble in alcohol may be responsible for the headache, in addition to ethanol. Tyramine, histamine (a vasodilator), and phenylethylamine are residual chemicals which have been identified in wine.[41] A large number of phenolic compounds are found in wines which are either derived from the grapes or produced via fermentation processes.[42] Parantainen[40] found several reasons to suggest that prostaglandins and related fatty acids may be involved in the generation of alcohol headaches inasmuch as they are important factors in other forms of headaches. Prostaglandin inhibitors effectively reduce some alcohol-related intolerance and other acute reactions as well as headache and other symptoms of a hangover.[40]

Ethyl alcohol is a lipid solvent, a scavenger of oxygen free radicals, and a membrane-active agent that has all the prerequisites for regulating the phospholipid-fatty acid-prostanoid cascade at several levels.[40] In addition, ethanol may be directly incorporated into these lipids to form ethyl esters with the fatty acids (such as arachidonic acid).

Tolfenamic acid (a phenolic compound) is a potent inhibitor of prostaglandin formation and action. This compound significantly reduced PGE_2 concentrations and the cardinal symptoms of hangover headache following the ingestion of alcohol.[43]

Ideal diets for migraine patients are not universally acceptable because of individual differences in chemical sensitivities. Raskin and Appenzeller[30] suggest that, in order to exclude diet as the cause of migraine, one should consume only the following foods for approximately 2 weeks: distilled water, lettuce, cauliflower, carrots, boiled or baked potatoes, cottage cheese, chicken, corn oil, olive oil, and distilled white vinegar. The premise is that if the headache were to persist on this diet, then diet should be ruled out as the causative factor. Such a diet, however, would activate headaches in many susceptible individuals because of sensitivities to the aromatic compounds in those natural foodstuffs.

REFERENCES

1. **Waters, W. E.,** *The Epidemiology of Migraine,* Boehringer Ingelheim, Bracknell-Berkshire, 1974, 69.
2. **Munro, J., Carini, C., Brostoff, J., and Zilkha, K.,** Food allergy in migraine, *Lancet,* 2, 13, 1980.
3. **Egger, J., Wilson, J., Carter, C. M., Turner, M. W., and Soothill, J. F.,** Is migraine food allergy?, *Lancet,* 2, 1, 1983.
4. **Grant, E. C. G.,** Food allergies and migraine, *Lancet,* 1, 966, 1979.
5. **Raskin, N. H.,** Pharmacology of migraine, *Annu. Rev. Pharmacol. Toxicol.,* 21, 463, 1981.
6. **Anthony, M., Hinterberger, H., and Lance, J. W.,** Plasma serotonin in migraine and stress, *Arch. Neurol.,* 16, 544, 1967.
7. **Curran, D. A., Hinterberger, H., and Lance, J. W.,** Total plasma serotonin, 5-hydroxyindole acetic acid and p-hydroxy-m-methoxymandelic acid excretion in normal and migrainous subjects, *Brain,* 88, 997, 1965.
8. **Jones, R. J., Amess, J. A. L., and Wachowicz, B.,** Migraine: a platelet disorder, *Lancet,* 2, 720, 1981.
9. **Dalsgaard, T. and Genefke, I. K.,** Serotonin release and uptake in platelets from healthy persons and migrainous patients in attack-free intervals, *Headache,* 14, 26, 1974.
10. **Hanington, E.,** Preliminary report on tyramine headache, *Br. Med. J.,* 2, 550, 1967.
11. **Boullin, D. J.,** Aetiology — clinical aspects, in *Cerebral Vasospasm,* Boullin, D. J. and Blaso, W. P., Eds., John Wiley & Sons, New York, 1980, 143.
12. **Schaechter, J. D. and Wurtman, R. J.,** Serotonin release varies with brain tryptophan levels, *Brain Res.,* 532, 203, 1990.
13. **Edwards, S. T. and Sandine, W. E.,** Public health significance of amines in cheese, *J. Dairy Sci.,* 64, 2431, 1981.
14. **Udenfriend, S., Lovenberg, W., and Sjoerdsma, A.,** Physiologically active amines in common fruits and vegetables, *Arch. Biochem. Biophys.,* 85, 487, 1959.
15. **Henderson, W. R. and Raskin, N. H.,** "Hot-dog" headache: individual susceptibility to nitrite, *Lancet,* 2, 1162, 1972.
16. **Baraboi, V. A.,** Effects of sodium gallate and some other polyphenols on adrenaline toxicity, *Fiziol. Zh. (Kiev),* 13, 809, 1967.
17. **Dorris, R. L. and Dill, R. E.,** Inhibition of catechol O-methyltransferase by n-butyl gallate, *Neuropharmacology,* 16, 631, 1977.
18. **Holmes, S. W.,** The spontaneous release of prostaglandins into the cerebral ventricles of the dog and the effect of external factors on the release, *Br. J. Pharmacol.,* 38, 653, 1970.
19. **Little, C. H., Stewart, A. G., and Fennessy, M. R.,** Platelet serotonin release in rheumatoid arthritis: a study in food intolerant patients, *Lancet,* 2, 297, 1983.
20. **Friberg, L., Olesen, J., Iversen, H. K., and Sperling, B.,** Migraine pain associated with middle cerebral artery dilatation: reversal by sumatriptan, *Lancet,* 338, 13, 1991.
21. **Olesen, J. and Edvinsson, L.,** Migraine: a research field matured for the basic neurosciences, *Trends Neurosci.,* 14, 3, 1991.
22. **Ekbom, K., et al.,** Treatment of acute cluster headache with sumatriptan, *N. Engl. J. Med.,* 325, 322, 1991.

23. **Ferrari, M. D., et al.,** Treatment of migraine attacks with sumatriptan, *N. Engl. J. Med.*, 325, 316, 1991.
24. **Raskin, N. H.,** Serotonin receptors and headache (editorial), *N. Engl. J. Med.*, 325, 353, 1991.
25. **Satel, S. L. and Gawin, F. H.,** Migrainelike headache and cocaine use, *J.A.M.A.*, 261, 2995, 1989.
26. **Cunningham, K. A. and Lakoski, J. M.,** Electrophysiological effects of cocaine and procaine in dorsal raphe serotonin neurons, *Eur. J. Pharmacol.*, 148, 457, 1988.
27. **Moore, P. K.,** The cardiovascular system, in *Prostanoids: Pharmacological, Physiological and Clinical Relevance*, Cambridge University Press, New York, 1985, chap. 3.
28. **Burnstock, G.,** Pathophysiology of migraine: a new hypothesis, *Lancet*, 1, 1397, 1981.
29. **Dalessio, D. J.,** Dietary migraine, *Am. Fam. Physician*, 6(6), 60, 1972.
30. **Raskin, N. H. and Appenzeller, O.,** Headache, in *Major Problems in Internal Medicine*, Vol. 19, Smith, L. D., Ed., W. B. Saunders, Philadelphia, 1980.
31. **Reif-Lehrer, L.,** A questionnaire study of the prevalence of Chinese Restaurant Syndrome, *Fed. Proc.*, 36(5), 617, 1977.
32. **Ghadimi, H., Kumar, S., and Abaci, F.,** Studies on monosodium glutamate ingestion. I. Biochemical exploration of Chinese Restaurant Syndrome, *Biochem Med.*, 5, 447, 1971.
33. **Ghadimi, H. and Kumar, S.,** Current status of monosodium glutamate, *Am. J. Clin. Nutr.*, 25, 643, 1972.
34. **Mansfield, L. E.,** The role of food allergy in migraine: a review, *Ann. Allergy*, 58, 313, 1987.
35. **Kandela, P.,** Migraine (a conference summary), *Lancet*, 2, 167, 1985.
36. **Markley, H. G.,** Verapamil and migraine prophylaxis: mechanisms and efficacy, *Am. J. Med.*, 90 (Suppl. 5A), 48S, 1991.
37. **Byer, J. A. and Dexter, J. D.,** Hypoglycemic migraine, *Mo. Med.*, 72, 194, 1974.
38. **Dalessio, D. J.,** Dietary migraine, *Am. Fam. Physician*, 6(6), 60, 1972.
39. **Raskin, N. H.,** Migraine, *West. J. Med.*, 123, 211, 1975.
40. **Parantainen, J.,** Possible roles of membrane lipids and prostaglandins in alcohol-related headache, *Med. Biol.*, 62, 1, 1984.
41. **Brainard, J. B.,** *Control of Migraine*, W. W. Norton, New York, 1979.
42. **Singelton, V. L. and Esau, P.,** *Phenolic Compounds in Grapes and Wine and Their Significance*, Academic Press, New York, 1969.
43. **Kaivola, S., Parantainen, J., Osterman, T., and Timonen, H.,** Hangover headache and prostaglandins: prophylactic treatment with tolfenamic acid, *Cephalalgia*, 3, 31, 1983.

13

Dermatitis

Gallates (Figure 13.1) have been identified in the causation of contact dermatitis in bakers and other workers handling these substances.[1] Vanillin is also a well-recognized sensitizer in contact dermatitis.[2] Contact dermatitis arising from contact with certain plants has been attributed to phenolic compounds. Urushiol (Figure 13.1) is a potent sensitizer found in poison oak/ivy.[3] The active contact allergen of *Phacelia crenulata* or "desert heliotrope" is geranylhydroquinone (Figure 13.1) and that of *Phacelia minor* is geranylgeranylhydroquinone.[3] Dermatitis-inducing furanocoumarins (psoralens; Figure 13.1) have been detected on leaf surfaces of eight species of plants belonging to the citrus and parsley families.[4]

The hairs of the stinging nettle are unique because of a combination of acetylcholine and histamine in the hair fluid.[5] The concentration of acetylcholine has been estimated to be greater than 50 m*M* (10 mg/ml).[6] Histamine is present at less than one tenth of this concentration.[5] Intradermal injection in man of a mixture of the two chemicals will imitate the sting produced by the hairs; however, neither alone will produce the response.[5]

Chronic urticaria has been elicited in some subjects challenged with aspirin, artificial food colors (such as tartrazine), and benzoates.[7-10] In a scatter diagram, Doeglas[7] identified a significant clustering of aspirin intolerance with food additive intolerance. Supramanian and Warner[9] have concluded that food-additive intolerance resulting in urticaria is not associated with atopy, suggesting that it is not an IgE-mediated phenomenon inasmuch as IgE antibodies to food additives have not been shown in human subjects. They were unable to offer a mechanism for additive intolerance. Artificial additives are most commonly phenolic (aromatic) compounds which may have the same pharmacological effects in increasing eicosanoid production in susceptible individuals as phenolic compounds in natural foods, as described in Chapter 1.

Photodermatitis in grocery store workers following contact with celery has been attributed to linear furanocoumarins.[11] Substances identified included psoralen, xanthotoxin, and bergapten.[11] An exposure of 18 μg/g fresh weight of the linear furanocoumarins was adequate to cause the contact dermatitis response.[12]

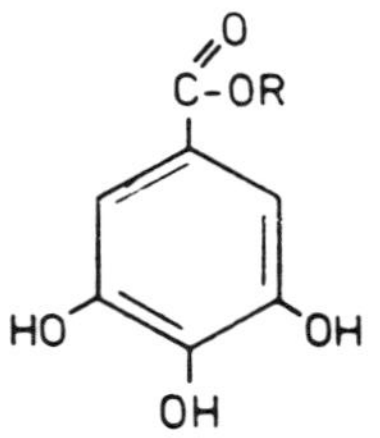

C-OR
O
HO
OH
OH

Structural formula of the gallates, in which R = C_3H_7 (propyl gallate), C_8H_{17} (octyl gallate) of $C_{12}H_{25}$ (dodecylgallate).

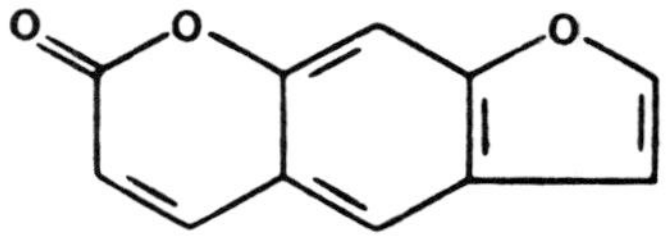

O
O
O

Psoralen

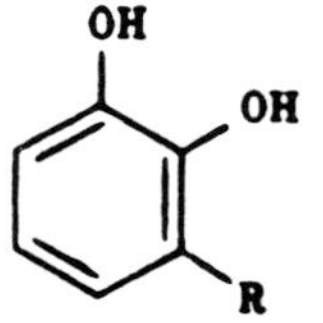

OH
OH
R

Urushiol

I R = $(CH_2)_{14}CH_3$
II R = $(CH_2)_7CH{=}{=}CH(CH_2)_5CH_3$
III R = $(CH_2)_7CH{=}{=}CHCH_2CH{=}{=}CH(CH_2)_2CH_3$
IV R = $(CH_2)_7CH{=}{=}CHCH_2CH{=}{=}CHCH{=}{=}CHCH_3$
V R = $(CH_2)_7CH{=}{=}CHCH_2CH{=}{=}CHCH_2CH=CH_2$

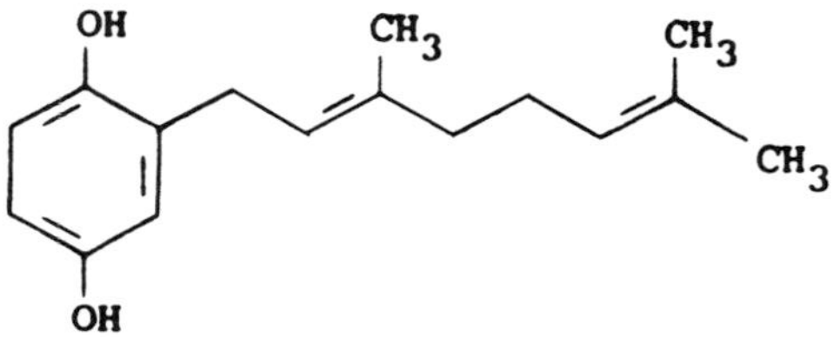

OH
CH_3
CH_3
CH_3
OH

Geranylhydroquinone

Figure 13.1 Phenolic compounds identified as agents capable of causing dermatitis.

Eicosanoid levels have been measured in the skin of adult patients with atopic dermatitis.[13,14] The inflammatory mediators prostaglandin E_2 (PGE_2) and leukotriene B_4 (LTB_4) were present in lesional skin of atopic subjects in biologically active concentrations. This led to the conclusion that since PGE_2 and LTB_4 are mediators able to induce cutaneous inflammation and to modulate cellular immunity, they may be involved in the biochemical process leading to atopic dermatitis.[13,14] After observing elevated skin levels of LTB_4 in patients with active atopic dermatitis, Thorsen et al.[14] proposed that drugs inhibiting relevant enzymes or blocking leukotriene receptors, when available for clinical use, might exhibit the beneficial effect of glucocorticosteroids without harboring the well-known steroidal side effects.

Histamine causes pruritus, erythema, a circumferential flare, and a central wheal upon subcutaneous injection.[15] This response is entirely blocked by a combination of H_1 and H_2 antagonists.[15] Prostaglandins and leukotrienes may be the mediators of the wheal and flare since they cause this effect when injected intradermally into human skin.[15-18] PGE_2 can amplify the wheal-and-flare reaction induced by intradermal injection of LTB_4 into the skin.[18]

An interesting study of mediators of erythema in human skin during the immediate and late-phase response was determined in ragweed- and grass-allergic patients with a blister-chamber technique.[19] The initial (immediate) erythema following injection of allergen was associated with a sharp rise in histamine levels at the first hour which progressively declined during the next 4 hours by 75%. LTC_4 levels were significantly elevated 4 to 6 hours after challenge. PGD_2 rose gradually to a peak at 5 to 6 hours. The conclusion from this study was that early rises in histamine were temporarily related to the immediate erythema, whereas the eicosanoids appearing in the skin after allergen challenge followed kinetics that corresponded to the time course of the late-phase response.

Stress and anxiety may trigger outbreaks of psoriasis.[20] The mechanism of this response seems likely to be mediated by increased levels of norepinephrine with subsequent liberation of arachidonic acid for synthesis of eicosanoids (see Chapter 1 for discussion of biochemistry).

Two cases of atopic eczema have been reported following oral and intravenous supplements of calcium given to children with milk and other food intolerances.[21] The supplements were not given in excess of the recommended daily allowances. There was a remission of symptoms when supplementation was discontinued in each case. Reasons for such a response were puzzling to those making the observation.[21] They did suggest that the calcium ion was possibly directly responsible for the adverse reaction. This is very possible when one considers that Ca^{2+}, upon binding with calmodulin, activates one or more lipases to release arachidonic acid for increased synthesis of eicosanoids. Verification of the role of calcium in urticaria has come from a study using nifedipine, a pure calcium channel blocker.[22] Patients who were refractory to maximally tolerated doses of antagonists to the H_1 and H_2 histamine receptors were studied in a double-blind crossover trial. The calcium channel blocker provided significant therapeutic efficacy for the patients with severe chronic idiopathic urticaria. For example, the hive index after 4 weeks of nifedipine was 0.003 vs. 0.736 after 4 weeks of placebo. This group at Baylor University's College of Medicine was also uncertain as to the reason for the effects of using nifedipine but suggested as possibilities (1) inhibition of histamine release, (2) inhibition of leukotriene release or function, or (3) inhibition of release or function of platelet activating factor.

REFERENCES

1. Joint FAO/WHO Expert Committee on Food Additives, Toxicological evaluation of certain food additives, *W.H.O. Food Addit. Ser.*, 10, 45, 1976.
2. **Cronin, E.,** *Contact Dermatitis,* Churchill Livingstone, Edinburgh, 1980, 524.
3. **Reynolds, G., Epstein, W., Terry, D., and Rodriguez, E.,** A potent contact allergen of *Phacelia* (Hydrophyllaceae), *Contact Dermatitis,* 6, 272, 1980.
4. **Zobel, A. M. and Brown, S. A.,** Dermatitis-inducing furanocoumarins on leaf surfaces of eight species of Rutaceous and Umbelliferous plants, *J. Chem. Ecol.,* 16, 693, 1990.
5. **Barlow, R. B. and Dixon, R. O. D.,** Choline acetyltransferase in the nettle *Urtica dioica* L., *Biochem. J.,* 132, 15, 1973.
6. **Emmelin, N. and Feldberg, W.,** Pharmacologically active substances in the fluid of nettle hairs (Urtica urens), *J. Physiol. (London),* 106, 440, 1947.
7. **Doeglas, H. M. G.,** Chronic urticaria: intolerance for aspirin and food additives and relationship with atopy, *Br. J. Dermatol.,* 108, 108, 1983.
8. **Stevenson, D. D., Simon, R. A., Lumry, W. R., and Mathison, D. A.,** Adverse reactions to tartrazine, *J. Allergy Clin. Immunol.,* 78, 182, 1986.
9. **Supramanian, G. and Warner, J. O.,** Artificial food additive intolerance in patients with angio-oedema and urticaria, *Lancet,* 2, 907, 1986.
10. **Lockey, S. D.,** Allergic reactions due to F.D. and C. yellow No. 5 tartrazine, an aniline dye used as a coloring agent in various steroids, *Ann. Allergy,* 17, 719, 1959.
11. **Trumble, J. T., Millar, J. G., Ott, D. E., and Carson, W. C.,** Seasonal patterns and pesticidal effects on the phototoxic linear furanocoumarins in celery, *Apium graveolens* L., *J. Agric. Food Chem.,* 40, 1501, 1992.
12. **Austad, J. and Kavli, G.,** Phototoxic dermatitis caused by celery infected by *Sclerotinia sclerotiorum, Contact Dermatitis,* 9, 448, 1983.
13. **Fogh, K., Herlin, T., and Kragballe, K.,** Eicosanoids in skin of patients with atopic dermatitis: prostaglandin E_2 and leukotriene B_4 are present in biologically active concentrations, *J. Allergy Clin. Immunol.,* 83, 450, 1989.
14. **Thorsen, S., Fogh, K., Broby-Johansen, U. B., and Sondergaard, J.,** Leukotriene B_4 in atopic dermatitis: increased skin levels and altered sensitivity of peripheral blood T-cells, *Allergy,* 45, 457, 1990.

15. **Serafin, W. E. and Austen, K. F.,** Mediators of immediate hypersensitivity reactions, *N. Engl. J. Med.*, 317, 30, 1987.
16. **Lewis, R. A.,** The biologically active leukotrienes: biosynthesis, metabolism, receptors, functions, and pharmacology, *J. Clin. Invest.*, 73, 889, 1984.
17. **Snyder, D. W. and Fleisch, J. H.,** Leukotriene receptor antagonists as potential therapeutic agents, *Annu. Rev. Pharmacol. Toxicol.*, 29, 123, 1989.
18. **Archer, C. B., Page, C. P., Juhlin, L., Morley, J., and McDonald, D. M.,** Delayed synergism between leukotriene B_4 and prostaglandin E_2 in human skin, *Prostaglandins*, 33, 799, 1987.
19. **Reshef, A., Kagey-Sobotka, A., Adkinson, N. F., Lichtenstein, L. M., and Norman, P. S.,** The pattern and kinetics in human skin of erythema and mediators during the acute and late-phase response (LPR), *J. Allergy Clin. Immunol.*, 84, 678, 1989.
20. **Hale, E.,** Brushing off dandruff and other flaky afflictions — psoriasis, *F.D.A. Consumer*, 22, 28, 1988.
21. **Devlin, J. and David, T. J.,** Intolerance to oral and intravenous calcium supplements in atopic eczema, *J. R. Soc. Med.*, 83, 497, 1990.
22. **Huston, D. P.,** New approaches to the therapy of urticaria, *Insights Allergy*, 5, 13, 1990.

14

Phenolic Compounds as Anti-Inflammatory Agents

I. DOSE-RESPONSE RELATIONSHIPS

Phenolic compounds have been studied extensively because of their pharmacological properties, which may cause adverse physiological-pharmacological effects, as discussed in Chapter 1, or, when used in correct quantities may be beneficial in medical therapy.[1-4] Lands and Hanel[4] mention phenolic compounds as agents which activate cyclooxygenase activity (yielding prostaglandins and thromboxanes) at low concentrations and inhibit such activity at high concentrations. This seemed paradoxical to them since the response could be either stimulatory or inhibitory, depending on the concentration of the phenolics or the environment of the cyclooxygenase. In addition, these scientists have directed attention to the fact that some phenolics have proven useful in modifying pathophysiologic conditions associated with the overproduction of prostaglandins.[4] Prostaglandin H synthase (PHS) is the initial enzyme in the cyclooxygenase pathway which commits arachidonic acid to the formation of prostaglandins (Figure 14.1).[5] Two catalytic activities follow PHS — cyclooxygenase and peroxidase (Figure 14.1). Phenol enhances the specific activity of PHS by approximately threefold; however, phenol inhibits the PHS cyclooxygenase at higher concentrations.[5] Acetaminophen (Tylenol®, Cetadol®, etc.; Figure 14.2) is a phenolic compound used clinically to inhibit cyclooxygenase activity, but dosage level is critical inasmuch as low doses given *in vivo* stimulate cyclooxygenase activity, which is typical of most phenolic compounds, whereas a higher dose level inhibits cyclooxygenase activity.[5] Phenol may be used as an example of concentration response, inasmuch as 200 μM phenol stimulates PHS activity whereas a concentration of 4500 μM inhibits activity.[6] A parallel effect of dose level has been reported in studies with serotonin.[7] Perfusion or incubation of nervous and vascular tissue with serotonin increases prostaglandin biosynthesis.[7] Conversely, high concentrations of serotonin result in a depression of cyclooxygenase activity.[7,8] For instance, cyclooxygenase activity was stimulated in a medium containing 10^{-7} to 10^{-5} M serotonin, but there was a

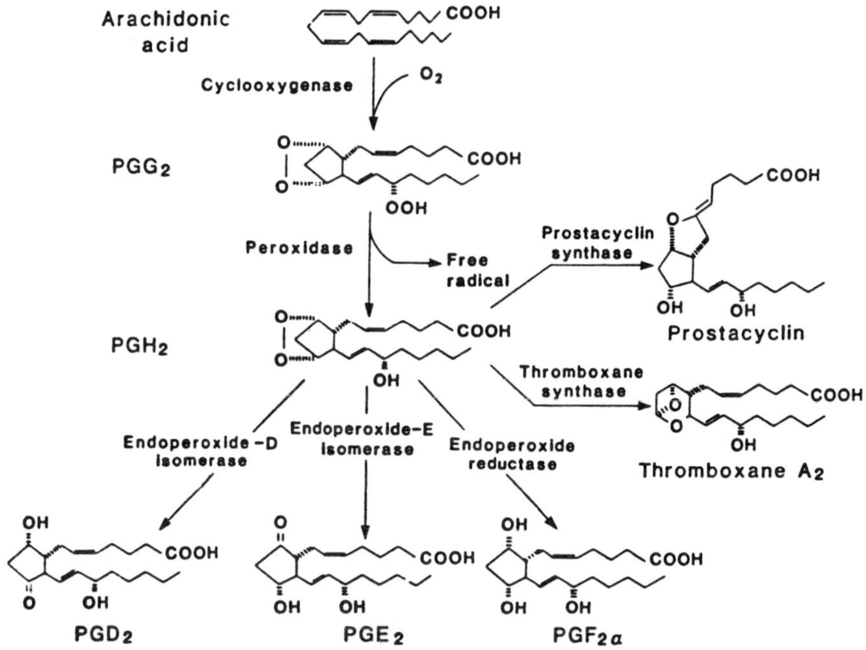

Figure 14.1 The metabolism of arachidonic acid to prostaglandins. (From Eling, T. E., et al., *Annu. Rev. Pharmacol. Toxicol.*, 30, 1, 1990. With permission.)

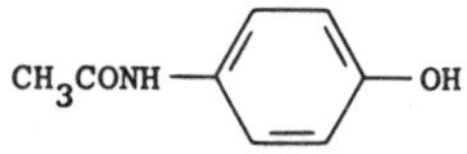

Figure 14.2 Acetaminophen.

significant decrease in cyclooxygenase activity upon incubation with 10^{-4} *M* serotonin.[7]

II. FLAVONOIDS IN THERAPY

Flavonoids in biology and medicine were the center of attention in symposia held in 1985 and 1987.[2,3] Biochemical, pharmacological, and structure-activity relationships of bioflavonoids and related phenolics have been studied extensively by scientists in many countries. A conclusion made from these studies is that all flavonoids, irrespective of the level of oxidation or stereochemistry of the heterocyclic ring, act as inhibitors of PHS provided they possess a catechol structure (adjacent hydroxyl groups) on at least one of the two aromatic rings (A or B; Figure 14.3).[9] Most of the flavonoids appear to be dual inhibitors of both cyclooxygenase and lipoxygenase pathways and to inhibit these enzymes in the cell with nearly the same potency.[10] Antioxidant properties of flavonoids are important in inhibitory

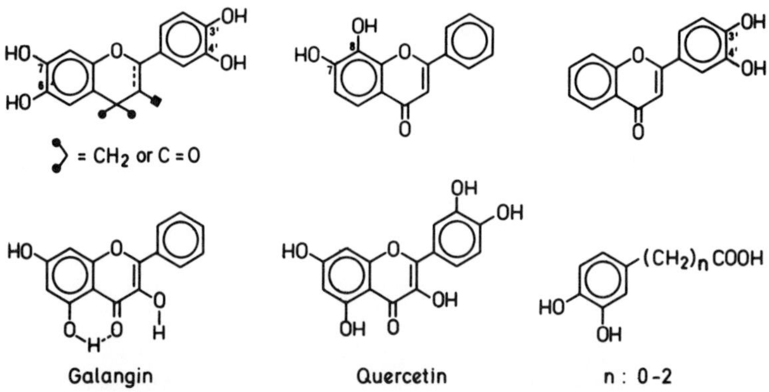

Figure 14.3 Prostaglandin inhibitors (substituted benzcatechin derivatives and their equivalents). (From Wagner, H., *Annu. Proc. Phytochem. Soc.*, 25, 409, 1985. With permission.)

responses of various oxidases.[11] Quercetin (a prominent flavonoid) is a potent inhibitor of lipoxygenase isolated from soybeans; however, this property was attributed to its effect on the structural specificity of the enzyme rather than to an antioxidant function.[12] The daily dietary intake of flavonoids is approximately 1 g, with vegetables being the primary source.[13] This level could activate cyclooxygenase activity in chemically sensitive individuals, considering the finding that concentrations of flavonols of 150 to 450 μM are lethal to enterocytes of the intestine, and this concentration may be achieved with ingestion of 1 g of flavonoids daily.[14]

III. OTHER PHENOLIC INHIBITORS OF CYCLOOXYGENASE AND LIPOXYGENASE PATHWAYS

Other phenolic compounds are also effective inhibitors of the cyclooxygenase and lipoxygenase pathways. For example, caffeic acid (a benzopyrene) is one of the most selective inhibitors reported for 5-lipoxygenase.[15] It inhibits 5-lipoxygenase in a noncompetitive manner.[15] Use of caffeic acid may in turn stimulate prostaglandin biosynthesis, according to some reports.[16] This could logically be due to shunting of arachidonic acid to the cyclooxygenase pathway once the lipoxygenase pathway is interrupted (Figure 14.4). Caffeoyl-β-phenethylamine stimulated prostaglandin biosynthesis at a lower concentration and inhibited it at a higher concentration,[16] as was discussed above relative to other phenolic compounds. The same effect was detected with derivatives of cinnamic acids and with caffeic acid without added groups.[16] Coniferyl alcohol (Figure 14.5), which lacks the carboxyl group, strongly inhibited prostaglandin biosynthesis even at a low concentration.[16]

Dewhirst[17] studied structure-activity relationships of phenolic compounds for inhibition of prostaglandin cyclooxygenase. A total of 63 phenolic

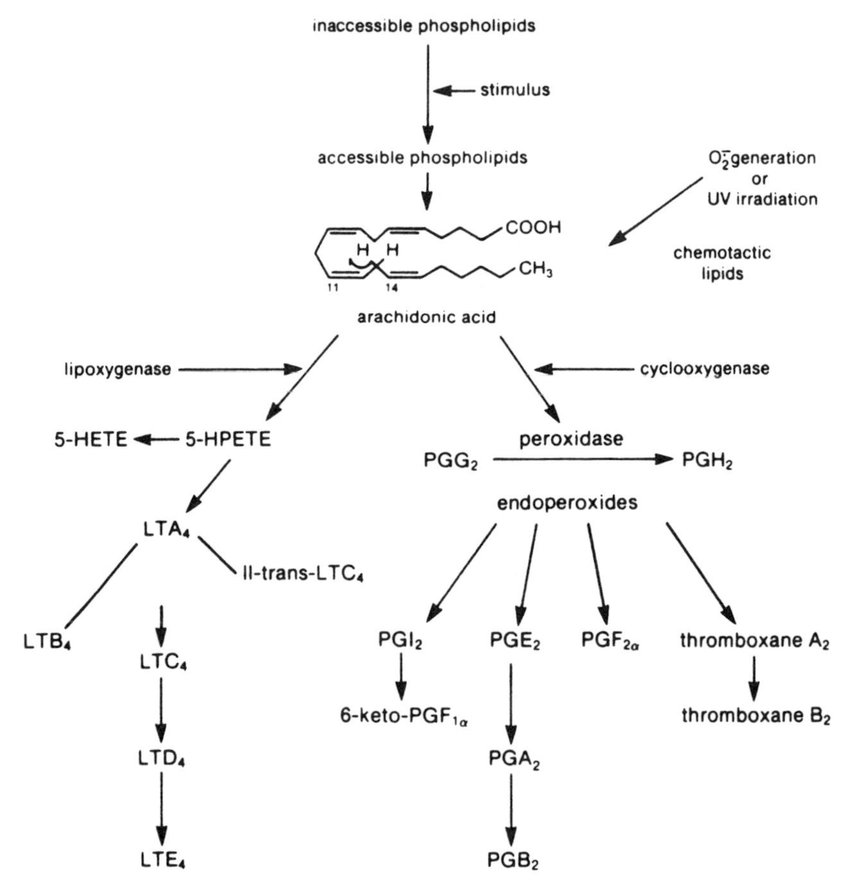

Figure 14.4 Major metabolites in the arachidonic cascade. (From Ninnemann, J. L., *Prostaglandins, Leukotrienes, and the Immune Response,* Cambridge University Press, New York, 1988, 13. With permission.)

$CH{=}CHCH_2OH$

OCH_3

OH

Figure 14.5 Coniferyl alcohol, an effective inhibitor of prostaglandin synthesis.

compounds were compared for their ability to inhibit sheep vesicular gland prostaglandin cyclooxygenase. Inhibition was increased by ring substituents which were electron donating and by substituents which were hydrophobic. Inhibition was decreased by steric masking of the phenolic hydroxyl. The most potent inhibitors possessed a two-aromatic-ring structure connected by a short bridge. In these inhibitors, one ring

FLAVONES

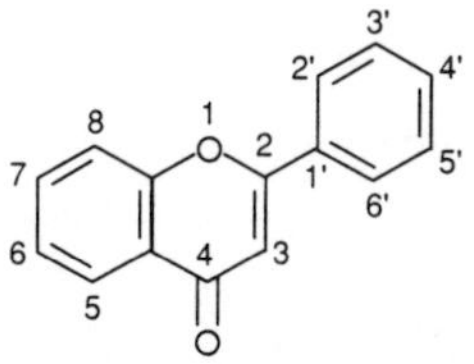

Compound	2'	3'	4'	5'	3	5	7
Flavone	-H	-H	-H	-H	-H	-H	-H
Kaempferol	-H	-H	-OH	-H	-OH	-OH	-OH
Fisetin	-H	-OH	-OH	-H	-OH	-OH	-OH
Morin	-OH	-H	-OH	-H	-OH	-OH	-OH
Myricetin	-H	-OH	-OH	-OH	-OH	-OH	-OH
Quercetin	-H	-OH	-OH	-H	-OH	-OH	-OH
Rutin	-H	-OH	-OH	-H	O-Rutinose	-OH	-OH
Cirsiliol	-H	-OH	-OH	-H	-H	-OH	$-OCH_3$
Pedalitin	-H	-OH	-OH	-H	-H	-OH	$-OCH_3$

Figure 14.6 Structure-activity relationships of flavones as inhibitors of the 5-lipoxygenase enzyme. Only flavones with hydroxyl groups at positions 4′, 3, and 7 were potent inhibitors. (From Welton, A. F., Hurley, J., and Will, P., *Plant Flavonoids in Biology and Medicine II: Biochemical, Cellular, and Medicinal Properties*, Alan R. Liss, New York, 1988, 301. With permission.)

was apolar and the other ring contained a phenolic hydroxyl ortho to the bridge; the bridge contained a Lewis base such that the compounds could form bidentate metal chelates. Welton et al.[10] made an additional observation of structural requirements of flavones and flavanones as inhibitors in the 5-lipoxygenase pathway. Only compounds which had hydroxyl groups at positions 4′, 3, and 7 (Figure 14.6) were potent inhibitors. A comprehensive listing of enzyme inhibitors and leukotriene receptor antagonists has been published by Fitzsimmons and Rokach.[18] These scientists have divided their listing into natural (e.g., flavonoids, caffeic acid, esculetin, etc.) and synthetic compounds (nordihydroguaiaretic acid [NDGA], eicosatetraynoic acid [ETYA], etc.).[18]

Esculetin (6,7-dihydroxycoumarin); (Figure 14.7) inhibits both 5- and 12-lipoxygenase, whereas the flavonoids are assumed to be more selective 5-lipoxygenase inhibitors.[19]

Eugenol (a flavor component in cloves; Figure 1.3) is a phenolic compound which inhibits PHS.[20] The mechanism of such inhibition was proposed by Thompson and Eling[20] to be due probably to competition with the arachidonic acid binding site on the enzyme. An additional conclusion

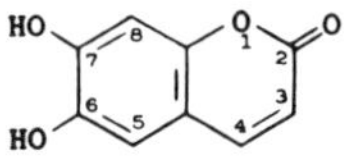

Figure 14.7 Esculetin (6,7-dihydroxycoumarin), an inhibitor of both 5- and 12-lipoxygenase.

from their study was that this inhibition occurs in addition to, or independent of, the effects on peroxide tone or peroxidase-dependent metabolism and applies to other phenolic compounds as well. Aspirin also inhibits cyclooxygenase activity, apparently due to aspirin acetylating a specific site on PHS.[21] This sterically inhibits arachidonate binding because of aspirin acetylating a serine at position 530 in the platelet enzyme.[22,23] Smith and Marnett[24] in a review paper explain that aspirin is effective because of its binding to the cyclooxygenase active site of prostaglandin endoperoxide (PGH) synthase (although it is a weak binding), thus competing with arachidonate for binding; following binding it causes acetylation of the protein. The acetylation is of a single serine residue.[24] This is an irreversible effect *in vivo* with no biochemical mechanism for hydrolyzing the acetylserine ester.[24]

IV. ANTIHISTAMINES

Certain flavonoids (quercetin, fisetin, apigenin, myricetin, phloretin, and kaemferol) inhibit antigen-induced histamine release (Figure 14.8) from sensitized mast cells.[24,25] Quercetin inhibits not only IgE-mediated allergic mediator release from mast cells, but also IgG-mediated histamine and peptidoleukotriene release from chopped lung fragments from actively sensitized guinea pigs.[26] After studying such flavonoids as quercetin, Welton et al.[27] observed that since flavonoids exhibit allergic mediator release activity and yet are selective inhibitors of the biosynthesis of proinflammatory arachidonic acid metabolites, they may be interesting prototypes which could lead to the discovery of very effective antiallergic and anti-inflammatory agents.

V. CALCIUM CHANNEL BLOCKERS

Another use of chemicals with benzene ring structures in therapy is observed in calcium-antagonist drugs such as verapamil, nifedipine, and diltiazem (Figure 14.9). These calcium-antagonist drugs principally interfere with the entry of calcium into cells through voltage-sensitive channels, whereas processes that involve the release of intracellular stores of calcium are relatively insensitive to these drugs.[28] High concentrations of certain calcium antagonists can also block the effects of norepinephrine competitively by an action at α-adrenergic receptors.[29] The reason for this effect, as discussed in Chapter 1, is that α-adrenergic receptor activation causes increased influx of Ca^{2+} and/or mobilizes intracellular Ca^{2+}, which

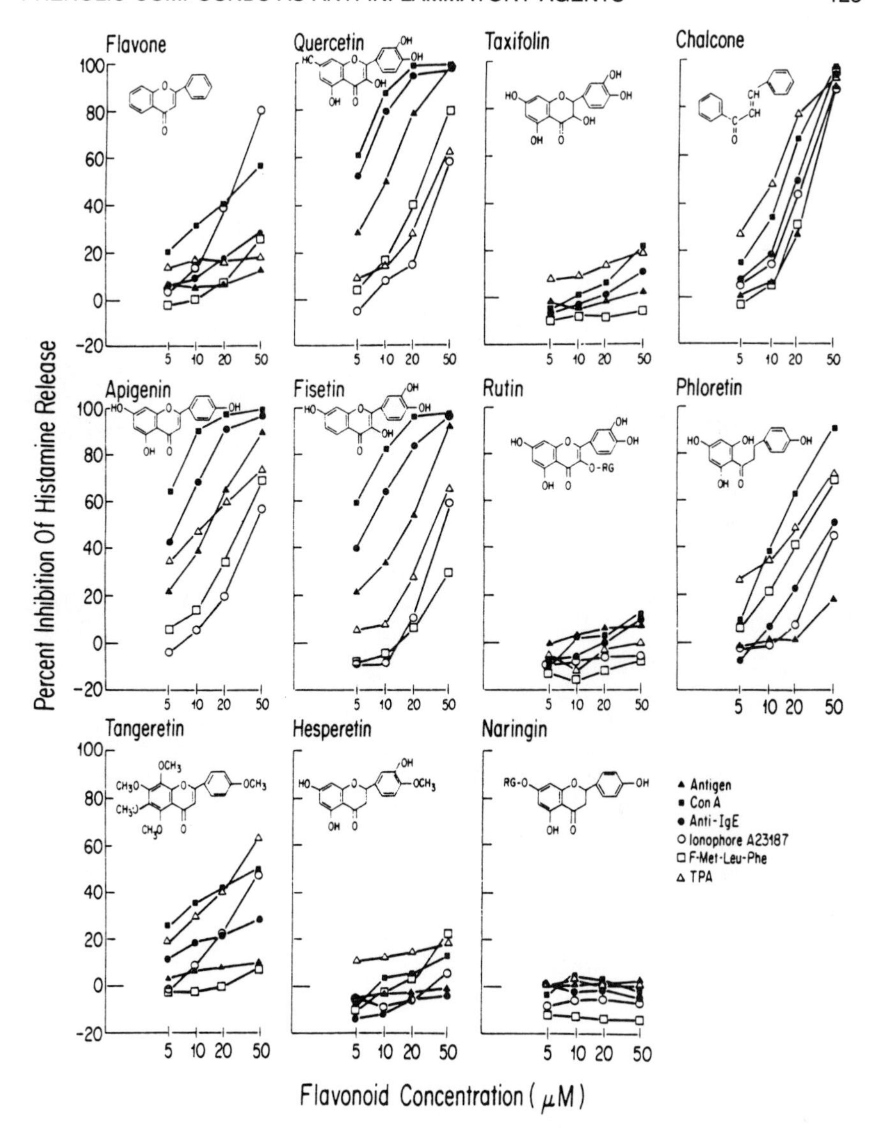

Figure 14.8 Structures and inhibitory activities of 11 flavonoids against basophil histamine release stimulated by 6 different secretogogues. Each point represents the averaged results of at least three experiments. (In the graph for taxifolin, the curve of ionophore A23187 virtually overlapped with the data points for antigen and was therefore omitted for clarity). RG = rhamnosylglucoside; ConA = concanavalin A; TPA = tetradecanoyl phorbal acetate. (From Middleton, E., Jr. and Drzewiecki, G., *Biochem. Pharmacol.*, 33, 3333, 1984. With permission.)

in turn binds with calmodulin and activates one or more lipases to release arachidonic acid for prostaglandin synthesis. Quercetin has also been mentioned as conceivably inhibiting *de novo* slow-reacting substance of

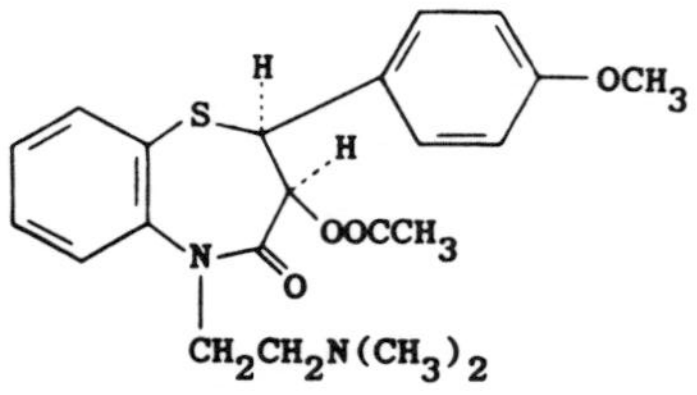

Diltiazem

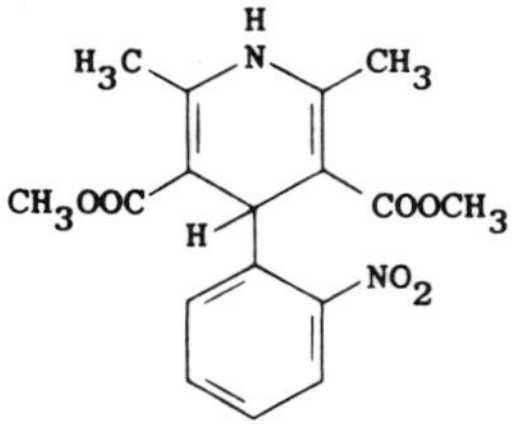

Nifedipine

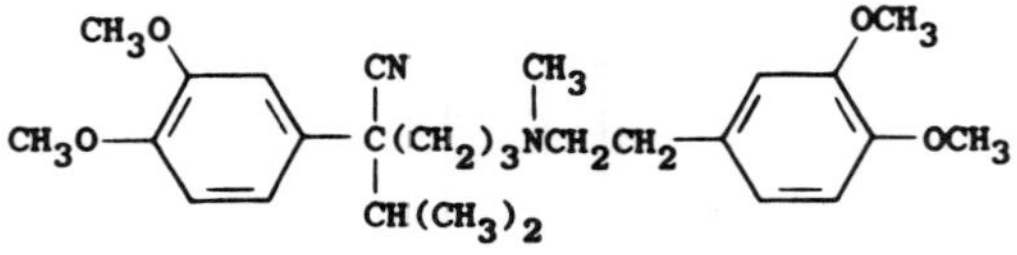

Verapamil

Figure 14.9 Calcium blocking agents.

anaphylaxis (SRS-A) biosynthesis by two mechanisms: (1) inhibition at the level of antigen-induced Ca^{2+} influx through the cell membrane, and (2) inhibition of a 5-lipoxygenase step after Ca^{2+} influx.[30]

Another calcium antagonist is the mineral Mg^{2+}.[31,32] Magnesium sulfate has been shown to be effective clinically as a bronchodilator,[32] acting on the airways in the same manner as nifedipine, which was cited above. Ample dietary magnesium is also needed to prevent inflammatory bowel disease.[33]

VI. GLUTATHIONE INTERACTIONS WITH AROMATIC COMPOUNDS

Glutathione (GSH) is considered to be one of the most important antioxidants and anticarcinogens in mammalian cells.[34] GSH is apparently synthesized in all mammalian cells at millimolar concentrations.[34] Certain chemicals, such as acetaminophen (Figure 14.2), quinones, and naphthoquinones (for example, menadione [vitamin K_3]), cause a depletion of GSH.[34,35] Quinones and naphthoquinones are natural pigment components of plants.[36] The reaction of GSH with menadione results in a menadione-GSH conjugate, thus causing a depletion of GSH available for cellular protection.[35] Additional dietary GSH has been suggested as of possible therapeutic value to replenish stores of this compound in these instances.[34]

Benzyl and allyl isothiocyanates found in vegetables and spices also conjugate with glutathione.[37] Both types of conjugates have been shown to cause considerable toxicity in RL-4 rat hepatocytes.[37] However, addition of excess GSH (>4 m*M*) to the free isothiocyanates, as well as their conjugates, abolished cytotoxicity up to the highest concentration tested (250 μ*M*).[37] Bruggeman et al.[37] suggest that the reason for this effect is that GSH conjugates are not able to enter cells; thus, the release of free isothiocyanate causes toxicosis, and an excess of GSH prevents formation of any appreciable amount of the free isothiocyanates.

GSH is involved in an important biochemical role as glutathione peroxidase.[38] Selenium is part of the active site of this peroxidase-destroying enzyme.[38] Selenium also regulates the balance between the proaggregatory and vasoconstrictory metabolite thromboxane A_2 (TxA_2) and the antiaggregatory and vasodilatatory prostacyclin PGI_2 in the cyclooxygenase pathway.[38] This essential element should be used with caution because of toxic effects when ingested in excessive amounts. Additional attention is advised in human nutrition to a study of possible competitive interaction between thiol (sulfur) compounds and selenium in the diet.

Butylated hydroxytoluene (BHT) and butylated hydroxyanethole (BHA) are antioxidants used extensively in foods and petroleum products.[39] BHT has toxic effects in lungs and livers of mice which are accentuated if there is a depression in GSH levels, inasmuch as BHT is detoxified by conjugation with GSH.[39]

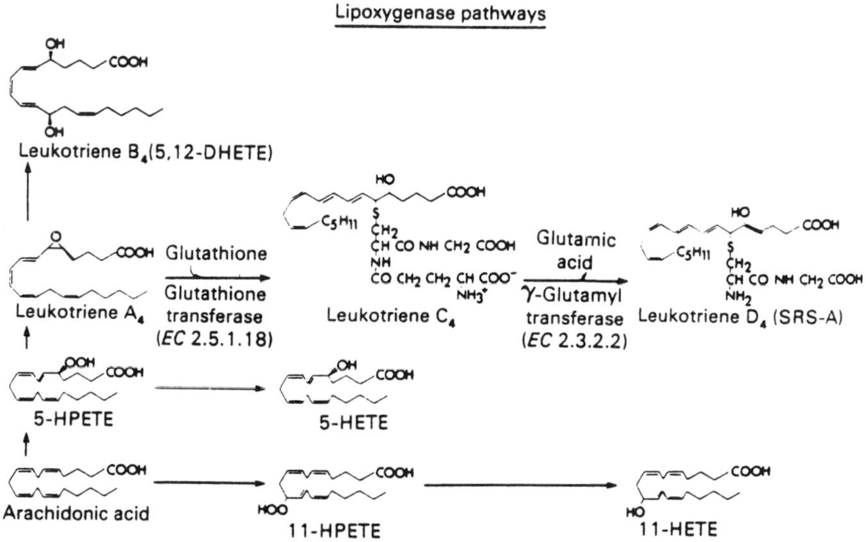

Lipoxygenase pathways

Leukotriene B_4(5,12-DHETE)

Glutathione → Glutathione transferase (*EC* 2.5.1.18)

Leukotriene A_4

CH_2 CH CO NH CH_2 COOH NH CO CH_2 CH_2 CH COO^- NH_3^+

Leukotriene C_4

Glutamic acid → γ-Glutamyl transferase (*EC* 2.3.2.2)

CH_2 CH CO NH CH_2 COOH NH_2

Leukotriene D_4 (SRS-A)

5-HPETE → 5-HETE

Arachidonic acid → 11-HPETE → 11-HETE

Figure 14.10 Diagram of the lipoxygenase pathways. Note the transfer of glutathione to LTA_4 to create LTC_4.

Vitamin E, vitamin C, and GSH appear to work in concert to inhibit lipid peroxidation.[40] Vitamin E functions as an antioxidant and undergoes oxidation to prevent the oxidation of polyunsaturated fatty acids.[41,42] Reddy et al.[43,44] and Maellaro et al.[45] have observed that GSH is only effective in inhibiting lipid peroxidation of liver microsomes which contain adequate amounts of vitamin E. Vitamin C, in conjunction with GSH, acts to regenerate vitamin E from its oxidized form.[46,47] Frei et al.[47] observed that ascorbic acid is more effective as an antioxidant than tocopherol in a plasma *in vitro* system. The effectiveness of ascorbic acid in controlling serum GSH levels was demonstrated in eight healthy men by Henning et al.[48] After 60 days of low ascorbic acid intake, the total GSH concentration and reduced GSH:oxidized GSH ratio were decreased in plasma. At the same time, NAD and NADP levels in RBCs were elevated.

Caution is mandatory in using GSH inasmuch as it may become affixed to leukotriene A_4 (LTA_4) via glutathione transferase to form LTC_4 (Figure 14.10). Subsequent loss of glutamic acid from GSH yields LTD_4 (SRS-A; Figure 14.10). Accordingly, the lipoxygenase pathway would have to be blocked simultaneously with the addition of GSH if chemical sensitivities exist in a patient.

Vitamin E, a cyclic compound, has been identified as an antioxidant on a physiological basis.[49,50] Vitamin E has also been designated as a surfactant, and high concentrations of vitamin E may function as a detergent.[49] Vitamin E modulates the biosynthetic pathways in arachidonic acid metabolism such as inhibiting fatty acid release and lipoxygenase activity (Figure 14.11).[49] Prostaglandin E_2 (PGE_2) synthesis in all brain regions of young and old mice except the cerebrum were significantly influenced by

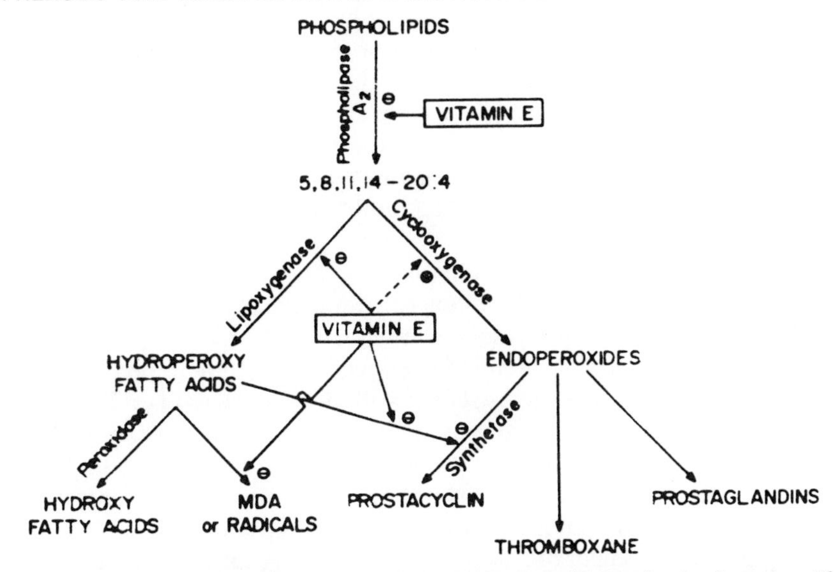

Figure 14.11 A model for the effects of vitamin E on the biosynthesis of prostanoids and hydroxy fatty acids from arachidonic acid. MDA = malondialdehyde (a lipid peroxide breakdown product). (From Panganamala, R. V. and Cornwell, D. G., *Ann. N.Y. Acad. Sci.*, 393, 376, 1982. With permission.)

dietary levels of vitamin E in a study by Meydani et al.[50] Diets low in vitamin E increased the production of PGE_2, whereas supplemental vitamin E caused a decrease in PGE_2.

The supplementation of 1600 IU of vitamin E to the diets of six healthy male volunteers per day substantially depressed thromboxane biosynthesis.[51] However, prostacyclin biosynthesis remained intact or was even enhanced in some individuals. The same responses were observed in a study with rats fed diets containing 5 and 15% corn oil in connection with three levels of vitamin E.[52] There was a stimulatory effect on PGI synthesis in the rat aorta concomitant with a decreased platelet TxB_2 production.

VII. THROMBOXANE SYNTHASE INHIBITORS

Aspirin irreversibly acetylates platelet cyclooxygenase and thereby inhibits production of TxA_2, a potent vasoconstrictor and platelet agonist.[53]

VIII. NEUTRALIZING DOSE LEVELS

The actual dosage of a phenolic compound which will arrest, rather than activate, a reaction falls in a narrow range. The provocative-neutralizing (symptom suppression) technique is very useful in finding the ideal dosage. This technique was discovered in 1957 by Carleton H. Lee of St. Joseph, Missouri, while he was attempting to find a better method of diagnosing and treating food allergy.[54] Lee began by studying variations in wheal sizes produced by intracutaneous injections of different dilutions of a given food extract. He soon learned that he was provoking symptoms with certain dilutions and relieving symptoms with other dilutions of the same extract. Later he learned he could relieve the symptoms of his patients by giving the previously determined relieving or neutralizing dilution.[54] Miller[54] and others adopted this principle and extended it not only to food extracts, but also to added pollen extracts, food additives, and chemical toxins to which their patients were exposed. The methods of administration also have been extended to sublingual and repository approaches.

Antigen preparations were initially either 1:5 or 1:10 of a food or pollen extract, or such a dilution series started with a 1% solution of an individual chemical. Patients react to either an underdose or an overdose of these antigen preparations. In general, the more sensitive the patient is to a particular chemical, the more dilute the initial neutralizing dose. After repeated application, the dosage required to neutralize becomes more concentrated, suggesting the development of a tolerance or physiological adaptation.

An explanation for the beneficial effects of neutralizing doses of food extracts is hypothesized by the author of this text to be due to correct dosage levels of phenolic compounds contained in the natural foods. As a consequence, there is adequate control of the cyclooxygenase and lipoxygenase pathways and the calcium channels. Supporting evidence is the finding that a nonneutralizing dose of one food may be neutralized by determining an appropriate neutralizing dose of another food. Those foods need not be in the same botanical family, suggesting that proteins are not the active agents.

Harmonic neutralizing dose was a term coined by E. L. Binkley, Jr., to describe the fact that several neutralizing dilutions may sometimes be found for one antigen in a particular patient.[54] For instance, dilution #2 in a 1:10 dilution series may relieve symptoms as well as dilutions 6 and 9, but intermediate dilutions would cause symptoms. The antigens in this case were food extracts, not single chemicals. Since several phenolic compounds are normal constituents in those foods and are found in different concentrations, this phenomenon could be attributed to different phenolic compounds with the ability to inhibit either the cyclooxygenase or lipoxygenase enzymes at their neutralizing dose levels. This does not apply in the testing of single phenolic compounds in searching for the proper neutralizing dose.

The next finding which verified the aforementioned hypothesis of dose responses relative to phenolic compounds in foods came while using dilutions of reagent-grade phenolic compounds. Foods either were ingested or sublingual challenges were made to elicit reactions, followed by sublingual applications of phenolic compounds (at their neutralizing dosages) to ascertain which phenolic compounds were needed to suppress reactions. Literature reviews offered clues as to possible phenolic compounds in each foodstuff.

Parantainen et al.[55] studied the effects of catecholamines on the formation and inhibition of formation of leukotrienes and prostaglandins in human polymorphonuclear leukocytes (PMNs). They referred to phenolic compounds with catechol structures in their study (for example, caffeic acid and 6,7-dihydroxycoumarin). Their observation was that compounds with catechol structures seemed to regulate the formation of leukotrienes and prostaglandins in diametrically opposite directions. Figure 14.12 illustrates their finding that increasing concentrations of catechol compounds in most cases reduced the formation of LTB_4, but there was a simultaneous increase in PGE_2. They concluded that catecholamines have a marked regulatory effect on eicosanoid synthesis in PMNs, particularly on the LT/PG ratio, through a receptor-independent chemical mechanism. However, they assumed that merely reducing synthesis of leukotrienes would obviate pathophysiologic conditions of tissue anaphylaxis without considering the pathology associated with simultaneous overproduction of prostaglandins. Comparisons of the effects of μm dosage levels illustrate the importance of dosage levels required to control the synthesis of leukotrienes and prostaglandins simultaneously. This information is cited as an illustration of the concept of the neutralizing dose which must inhibit the formation of one eicosanoid without increasing the synthesis of another. Counterbalancing chemicals which each successfully inhibit prostaglandin and leukotriene synthesis is another solution to these pathological disturbances.

Dilutions of the phenolic compounds may be made starting with a 1% solution, and then dilutions of 1:10 or 1:5 are made from that. Dorothy Sudweeks, a chemist working with Dennis W. Remington, M.D., Provo, Utah, pioneered the novel idea of using a 0.005 *M* solution, followed by preparing 1:1 and 1:9 dilutions of the 0.005 *M* solution. The patient applies 1 drop of the 1:9 dilution sublingually, waits about 10 seconds, applies 1 drop of the 1:1 solution, waits, and then applies 3 drops of the 0.005 *M* solution (or 4 drops if the chemical is dissolved in ethanol). Use of molar solutions reduces the number of dilutions required to find a neutralizing dose. It also reduces the disparity in the number of molecules of a chemical in comparable dilutions when comparing small molecules with large ones (for example, phenol mol wt = 94.11; quercetin mol wt = 302.23).

Drops of the chemicals should be taken following ingestion or inhalation of offending foods, pollens, or other substances toxic to the patients in order to "neutralize" the adverse effects.

IX. DEVELOPMENT OF CHEMICAL TOLERANCE (DESENSITIZATION)

The ultimate goal of prophylaxis in dealing with chemical sensitivities is development of tolerance (desensitization, hyposensitivity). In the process of using the neutralizing dose described in the previous section, the dosage level needed to neutralize a reaction increases over time.

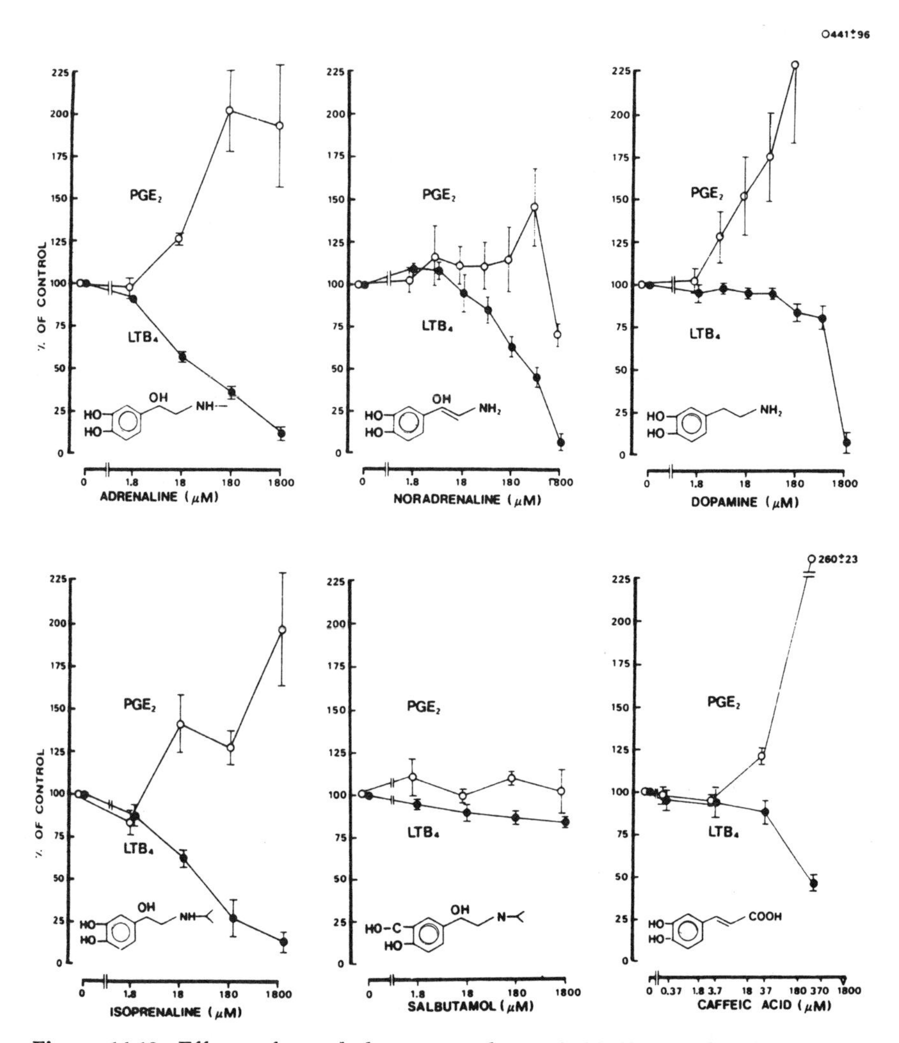

Figure 14.12 Effects of catechol compounds on A 23187-stimulated release of LTB_4 and PGE_2 from human PMNs. Baseline value for PGE_2 was 58 pg/10^6 cells and for LTB_4 8.1 ng/10^6 cells. The IC_{50} values, calculated from semilogarithmic dose-response curves, for LTB_4 release were as follows: adrenaline 46 ± 10, isoprenaline 73 ± 73, caffeic acid 370 ± 100, noradrenaline 419 ± 105, and dopamine 990 ± 95 μ*M* (mean ± SE, N = 5 to 9). Salbutamol had no effect. (From Parantainen, J., et al., *Biochem. Pharmacol.*, 40, 961, 1990. With permission.)

Initially, microgram amounts suffice, but after repeated sublingual applications the dosage needed to neutralize may be 5 mg, and eventually 125 mg or more (see Chapter 15 for author's findings). Degrees of adverse physiological responses to nonneutralizing doses begin to subside as an indication of the development of tolerance. Once tolerance has been attained, a question is raised as to the need for "booster doses" in the future. The answer must await the experiences of those who have pioneered this process.

The sublingual (oral) approach to desensitization is less hazardous than the classical, incremental subcutaneous desensitization protocol.[56-59] This is particularly true for patients with hypersensitivities who tend to have anaphylactic reactions induced by parenteral therapy.[58] No beneficial effect of oral administration of allergen to allergic patients was reported in some studies of oral immunotherapy.[59,60] However, Taudorf et al.[61] demonstrated a clinical effect of oral immunotherapy in birch pollinosis. The highest oral dosage level they used was about 200 times more than the dosage used in conventional subcutaneous immunotherapy. This was an 18-month study covering two birch pollen seasons. Enteric-coated capsules were used to bypass the stomach, and dosage levels were increased incrementally to 1.2×10^6 Biological Units. The study was double-blind and placebo controlled. Such treatment caused a decrease in eye symptom scores and conjunctival sensitivity to birch pollen, as determined by the conjunctival provocation test. There was also a highly significant decrease in skin reactivity to birch allergen preparation via skin prick tests. Gastrointestinal symptoms were the most frequent side effect, with abdominal swelling, pain, and/or diarrhea. This is not surprising, inasmuch as cross-sensitivities between birch pollen and apples, etc., have been reported (see Chapter 3). Fatigue was an associated side effect accompanying some of the gastrointestinal effects. As doses increased some subjects had side effects and were given lesser doses until they could tolerate higher levels. Thus, individual variations in sensitivity and the rate of developing tolerance are pertinent considerations in incremental dosage levels and time elements required to cause desensitization.

Induction of tolerance to individual allergenic chemicals has also come to fruition by sequential oral challenges.[62-69] Reduced responsiveness of adenylate cyclase to hormones and neurotransmitters is induced by chronic exposure to the agonist.[70,71] Desensitization is a common feature of β-adrenergic receptor systems, also due to down-regulation of the receptors by agonists.[71,72] There is a diminution in the number of β-receptors in the plasma membranes when the cells are exposed to β-adrenergic agonists either *in vivo* or *in vitro*.[72]

REFERENCES

1. **Walley, W. B.,** The toxicity of plant phenolics, in *The Pharmacology of Plant Phenolics,* Fairbairn, J. W., Ed., Academic Press, New York, 1959, 27.
2. **Cody, V., Middleton, E., Jr., and Harborne, J. B., Eds.,** *Plant Flavonoids In Biology and Medicine I: Biochemical, Pharmacological, and Structure-Activity Relationships,* Alan R. Liss, New York, 1986.

3. **Cody, V., Middleton, E., Jr., Harborne, J. B., and Beretz, A., Eds.,** *Plant Flavonoids in Biology and Medicine II: Biochemical, Cellular, and Medicinal Properties,* Alan R. Liss, New York, 1988.
4. **Lands, W. E. M. and Hanel, A. M.,** Inhibitors and activators of prostaglandin biosynthesis, in *Prostaglandins and Related Substances,* Pace-Asciak, C. and Granström, E., Eds., Elsevier, New York, 1983, 203.
5. **Eling, T. E., Thompson, D. C., Foureman, G. L., Curtis, J. F., and Hughes, M. F.,** Prostaglandin H synthase and xenobiotic oxidation, *Annu. Rev. Pharmacol. Toxicol.,* 30, 1, 1990.
6. **Egan, R. W., Gale, P. H., Beveridge, G. C., Marnett, L. J., and Kuehl, F. A., Jr.,** Direct and indirect involvement of radical scavengers during prostaglandin biosynthesis, *Adv. Prostaglandin Thromboxane Res.,* 6, 153, 1980.
7. **Dragan, Y. P. and Ellis, E. F.,** 5-Hydroxytryptamine and the metabolism of arachidonic acid by the lipoxygenase and cyclooxygenase of washed human platelets, *Biochem. Pharmacol.,* 40, 309, 1990.
8. **Harada, Y. and Kawamura, M. K. M.,** Is platelet 5-HT involved in prostaglandin endoperoxide generation as an endogenous tryptophan-like cofactor?, *Jpn. J. Pharmacol.,* 43(S), 88P, 1987.
9. **Wagner, H.,** New plant phenolics of pharmaceutical interest, in *Annu. Proc. Phytochemistry Society of Europe,* Van Sumere, C. F. and Lea, P. J., Eds., Oxford University Press, New York, 1985, 409.
10. **Welton, A. F., Hurley, J., and Will, P.,** Flavonoids and arachidonic acid metabolism, in *Plant Flavonoids in Biology and Medicine II: Biochemical, Cellular, and Medicinal Properties,* Alan R. Liss, New York, 1988, 301.
11. **Stafford, H. A.,** *Flavonoid Metabolism,* CRC Press, Boca Raton, FL, 1990, chap. 15.
12. **Ratty, A. K., Sunamoto, J., and Das, N. P.,** Interactions of flavonoids with 1,1-diphenyl-2-picrylhydrazyle free radical, liposomal membranes and soybean lipoxygenase-1, *Biochem. Pharmacol.,* 37, 989, 1988.
13. **Brown, J. P.,** A review of the genetic effects of naturally occurring flavonoids, anthraquinones and related compounds, *Mutat. Res.,* 75, 243, 1980.
14. **Canada, A. T., Watkins, W. D., and Nguyen, T. D.,** The toxicology of flavonoids to guinea pig enterocytes, *Toxicol. Appl. Pharmacol.,* 99, 357, 1989.
15. **Koshihara, Y., Neichi, T., Murota, S-I., Lao, A-N., Tujimoto, Y., and Tatsuno, T.,** Caffeic acid is a selective inhibitor for leukotriene biosynthesis, *Biochim. Biophys. Acta,* 792, 92, 1984.
16. **Tseng, C-F., Mikajri, A., Shibuya, M., Goda, Y., Elizuka, Y., Padmawinata, K., and Sankawa, U.,** Effect of some phenolics on the prostaglandin synthesizing enzyme system, *Chem. Pharmacol. Bull.,* 34, 1380, 1986.
17. **Dewhirst, F. E.,** Structure-activity relationships for inhibition of prostaglandin cyclooxygenase by phenolic compounds, *Prostaglandins,* 20, 209, 1980.
18. **Fitzsimmons, B. J. and Rokach, J.,** Enzyme inhibitors and leukotriene receptor antagonists, in *Leukotrienes and Lipoxygenases,* Rokach, J., Ed., Elsevier, New York, 1989, chap. 6.
19. **Stjernschantz, J.,** The leukotrienes, *Med. Biol.,* 62, 215, 1984.
20. **Thompson, D. and Eling T.,** Mechanism of inhibition of prostaglandin H synthase by eugenol and other phenolic peroxidase substrates, *Mol. Pharmacol.,* 36, 890, 1989.
21. **Roth, G. J., Stanford, N., and Majerus, P. W.,** Acetylation of prostaglandin synthase by aspirin, *Proc. Natl. Acad. Sci. U.S.A.,* 72, 3073, 1975.

22. **De Witt, D. L. and Smith, W. L.,** Primary structure of prostaglandin G/H synthase from sheep vesicular gland determined from the complementary DNA sequence, *Proc. Natl. Acad. Sci. U.S.A.*, 85, 1412, 1988.
23. **De Witt, D. L., El-harith, E. A., Kraemer, S. A., Yao, E. F., Armstrong, R. L., and Smith, W. L.,** The aspirin and heme-binding sites of ovine and murine PG endoperoxidase synthases, *J. Biol. Chem.*, 265, 5192, 1990.
24. **Smith, W. L. and Marnett, L. J.,** Prostaglandin endoperoxide synthase: structure and catalysis, *Biochim. Biophys. Acta*, 1083:1, 1991.
25. **Middleton, E., Jr. and Drzewiecki, G.,** Naturally occurring flavonoids and human basophil histamine release, *Int. Arch. Allergy Appl. Immun.*, 77, 155, 1985.
26. **Hope, W. C., Welton, A. F., Fiedler-Nagy, C., Batula-Bernardo, C., and Coffey, J. W.,** *In vitro* inhibition of the biosynthesis of slow reacting substance of anaphylaxis (SRS-A) and lipoxygenase activity by quercetin, *Biochem. Pharmacol.*, 32, 367, 1983.
27. **Welton, A. F., Tobias, L. D., Fiedler-Nagy, C., Anderson, W., Hope, W., Meyers, K., and Coffey, J. W.,** Effect of flavonoids on arachidonic acid metabolism, in *Plant Flavonoids in Biology and Medicine I: Biochemical, Pharmacological, and Structure-Activity Relationships*, Cody, V., Middleton, E., and Harborne, J. B., Eds., Alan R. Liss, New York, 1986, 231.
28. **Snyder, S. H. and Reynolds, I. J.,** Calcium-antagonist drugs: receptor interactions that clarify therapeutic effects, *New Engl. J. Med.*, 313, 995, 1985.
29. **Triggle, D. J.,** Biochemical pharmacology of calcium blockers, in *Calcium Blockers*, Flaim, S. F. and Zelis, R., Eds., Urban and Schwarzenberg, Baltimore, MD, 1982, 121.
30. **Fewtrell, C. M. S. and Gomperts, B. D.,** Quercetin: a novel inhibitor of Ca^{2+} influx and exocytosis in rat peritoneal mast cells, *Biochim. Biophys. Acta*, 469, 52, 1977.
31. **Levine, B. S. and Coburn, J. W.,** Magnesium, the mimic/antagonist of calcium, *N. Engl. J. Med.*, 310:1253, 1984.
32. **Lindeman, K. S., Hirshman, C. A., and Freed, A. N.,** Effect of magnesium sulfate on bronchoconstriction in the lung periphery, *J. Appl. Physiol.*, 66, 2527, 1989.
33. **Galland, L.,** Magnesium and inflammatory bowel disease, *Magnesium*, 7, 78, 1988.
34. **Hagen, T. M., Wierzbicka, T., Sillau, A. H., Bournan, B. B., and Jones, D. P.,** Bioavailability of dietary glutathione: effect on plasma concentration, *Am. J. Physiol.*, 259, G524, 1990.
35. **Takahashi, N., Schreiber, J., Fischer, V., and Mason, R. P.,** Formation of glutathione-conjugated semiquinones by the reaction of quinones with glutathione: an ESR study, *Arch. Biochem. Biophys.*, 252, 41, 1987.
36. **Robinson, T.,** *The Organic Constituents of Higher Plants — Their Chemistry and Interrelationships*, 4th ed., Cordus Press, North Amherst, MA, 1980, 58.
37. **Bruggeman, I. M., Temmink, J. H. M., and Van Bladeren, P. J.,** Glutathione- and cysteine-mediated cytoxicity of allyl and benzyl isothiocyanate, *Toxicol. Appl. Pharmacol.*, 83, 349, 1986.
38. **Neve, J.,** Physiological and nutritional importance of selenium, *Experientia*, 47, 187, 1991.
39. **Mizutani, T., Nomura, H., Nakanishi, K., and Fujita, S.,** Hepatotoxicity of butylated hydroxytoluene and its analogs in mice depleted of hepatic glutathione, *Toxicol. Appl. Pharmacol.*, 87, 166, 1987.

40. **Graham, K. S., Reddy, C. H., and Scholz, R. W.,** Reduced glutathione effects on α-tocopherol concentration of rat liver microsomes undergoing NADPH-dependent lipid peroxidation, *Lipids,* 24(11), 909, 1989.
41. **Dam, H.,** Influence of antioxidants and redox substances on signs of vitamin E deficiency, *Pharmacol. Rev.,* 9, 1, 1957.
42. **Machlin, L. J.,** The biological consequences of feeding polyunsaturated fatty acids to antioxidant-deficient animals, *J. Am. Oil Chem. Soc.,* 40, 368, 1963.
43. **Reddy, C. C., Scholz, R. W., Thomas, C. E., and Massaro, E. J.,** Vitamin E dependent glutathione inhibition of rat liver microsomal lipid peroxidation, *Life Sci.,* 31, 571, 1982.
44. **Reddy, C. C., Scholz, R. W., Thomas, C. E., and Massaro, E. J.,** Evidence for a possible protein-dependent regeneration of vitamin E in rat liver microsomes, *Ann. N.Y. Acad. Sci.,* 393, 193, 1982.
45. **Maellaro, E., Casini, A. F., Del Bello, B., and Comporti, M.,** Lipid peroxidation and antioxidant systems in the liver injury produced by glutathione depleting agents, *Biochem. Pharmacol.,* 39, 1513, 1990.
46. **Leedle, R. A. and Aust, S. D.,** Effect of glutathione on the vitamin E requirement for inhibition of liver microsomal lipid peroxidation, *Lipids,* 25, 241, 1990.
47. **Frei, B., England, L., and Ames, B. N.,** Ascorbate is an outstanding antioxidant in human blood plasma, *Proc. Natl. Acad. Sci. U.S.A.,* 86, 6377, 1989.
48. **Henning, S. M., Zhang, J. Z., McKee, R. W., Swendseid, M. E., and Jacob, R. A.,** Glutathione blood levels and other oxidant defense indices in men fed diets low in vitamin C, *J. Nutr.,* 121, 1969, 1991.
49. **Panganamala, R. V. and Cornwell, D. G.,** The effects of vitamin E on arachidonic acid metabolism, *Ann. N.Y. Acad. Sci.,* 393, 376, 1982.
50. **Meydani, S. N., Macauley, J. B., and Blumberg, J. R.,** Influence of dietary vitamin E and selenium on the ex-vivo synthesis of prostaglandin E_2 in brain regions of young and old rats, *Prostaglandins Leukotrienes Med.,* 18, 337, 1985.
51. **FitzGerald, G. A. and Brash, A. R.,** Endogenous prostacyclin and thromboxane biosynthesis during vitamin E therapy in man, *Ann. N.Y. Acad. Sci.,* 393, 209, 1982.
52. **Chan, A. C. and Hamelin, S. S.,** The effects of vitamin E and corn oil on prostacyclin and thromboxane B_2 synthesis in rats, *Ann. N.Y. Acad. Sci.,* 393, 201, 1982.
53. **Roth, G. J. and Majerus, P. W.,** The mechanism of the effect of aspirin on human platelets, *J. Clin. Invest.,* 56, 624, 1975.
54. **Miller, J. B.,** *Food Allergy — Provocative Testing and Injection Therapy,* Charles C Thomas, Springfield, IL, 1972, chap. 2.
55. **Parantainen, J., Alanko, J., Moilanen, E., Metsä-Ketelä, T., Asmawi, J., and Vapaatalo, H.,** Catecholamines inhibit leukotriene formation and decrease leukotriene/prostaglandin ratio, *Biochem. Pharmacol.,* 40, 961, 1990.
56. **Scadding, G. K. and Brostoff, J.,** Low dose sublingual therapy in patients with allergic rhinitis due to house dust mite, *Clin. Allergy,* 16, 483, 1986.
57. **Platt-Mills, T. A. E.,** Oral immunotherapy: a way forward? (Editorial), *J. Allergy Clin. Immunol.,* 80, 129, 1987.
58. **Stark, B. J., Wendel, G. D., and Sullivan, T. J.,** Oral desensitization for penicillin sensitivity, *J.A.M.A.,* 257, 1474, 1987.

59. **Taudorf, E., Laursen, L. C., Djurup, R., Kappelgaard, E., Pedersen, C. T., Soborg, M., Wilkinson, P., and Weeke, B.,** Oral administration of grass pollen to hay fever patients: An efficacy study of oral hyposensitization, *Allergy,* 40, 321, 1985.
60. **Cooper, P. J., Darbyshire, J., Nunn, A. J., and Warner, J. O.,** A controlled trial of oral hyposenstization in pollen asthma and rhinitis in children, *Clin. Allergy,* 14, 1541, 1984.
61. **Taudorf, E., Laursen, L. C., Lanner, A., Björksten, B., Dreborg, S., Sobørg, M., and Weeke, B.,** Oral immunotherapy in birch pollen hay fever, *J. Allergy Clin. Immunol.,* 80, 153, 1987.
62. **Michel, O., Naeije, N., Bracamonte, M., Duchateau, J., and Sergysels, R.,** Decreased sensitivity to tartrazine after aspirin desensitization in an asthmatic patient intolerant to both aspirin and tartrazine, *Ann. Allergy,* 52, 368, 1984.
63. **Chase, M. W.,** The induction of tolerance to allergenic chemicals, *Ann. N.Y. Acad. Sci.,* 392, 228, 1982.
64. **Kagnoff, M. F.,** Oral tolerance, *Ann. N.Y. Acad. Sci.,* 392, 248, 1982.
65. **Pleskow, W. W., Stevenson, D. D., Mathison, D. A., Simon, R. A., Schatz, M., and Zeiger, R. S.,** Aspirin desensitization in aspirin-sensitive asthmatic patients. Clinical manifestations and characterization of the refractory period, *J. Allergy Clin. Immunol.,* 69, 11, 1982.
66. **Stevenson, D. D.,** Aspirin sensitivity: new approaches to detection and desensitization, *Intern. Med.,* 8, 60, 1987.
67. **Stevenson, D. D.,** Proposed mechanisms of aspirin sensitivity reactions (editorial), *J. Allergy Clin. Immunol.,* 80, 788, 1987.
68. **Wedner, H. J.,** Reintroduction of drugs in the allergic patient, *Insights Allergy,* Vol. 3, No. 3, 1, 1988.
69. **Tanfin, Z. and Harbon, S.,** Heterologous regulations of cAMP responses in pregnant rat myometrium. Evolution from a stimulatory to an inhibitory prostaglandin E_2 and prostacyclin effect, *Mol. Pharmacol.,* 32, 249, 1987.
70. **Swillens, S., Boeynaems, J. M., and Dumont, J. E.,** Theoretical analysis of the cAMP cascade. 2. Theoretical considerations of the regulatory steps in the cAMP cascade system, *Methods Enzymol.,* 159, 25, 1988.
71. **Stiles, G. L., Caron, M. G., and Lefkowitz, R. J.,** β-Adrenergic receptors: biochemical mechanisms of physiological regulation, *Physiol. Rev.,* 64, 661, 1984.
72. **Stone, E. A.,** Central cyclic-AMP-linked noradrenergic receptors: new findings on properties as related to the actions of stress, *Neurosci. Biobehav. Rev.,* 11, 391, 1987.

15

Personal Findings of an Exploring Scientist

A preliminary history of findings and an introduction to concepts were included in the introduction to the text. The information which follows is a summary of concepts under investigation and the results of testing those concepts. Phenolic and other compounds found to be most effective in normalizing my body responses are listed with an explanation of the pharmacology involved and dosages needed.

The prevailing theory in arresting reactions using the neutralizing dose was that a reaction to any chemical could be controlled by using that chemical as its own control in a negative feedback response. Initially, I identified a total of approximately 30 aromatic, or cyclic, compounds which would neutralize reactions to 62 foods or food additives. With this information, I prepared a dot chart (Figure 15.1) which identified those chemicals most effective in arresting a reaction to specific foods. The assumption was that these specific chemicals were present in the foods tested. An important observation was that gallic acid, vanillylamine, and quercetin (a flavonol) were three chemicals which were effective in controlling reactions to numerous foods across many botanically unrelated food families (see dot chart). The reason for these responses became scientifically understandable in the 1980s due to greater knowledge of eicosanoids and the chemical nature of substances which block both the cyclooxygenase and lipoxygenase pathways, as well as calcium channels. With this new understanding, my attention then became focused on finding the biochemical pathways individual chemicals were affecting, followed by a survey to learn which chemicals might be duplicating each other in activating and/or suppressing those pathways. This then narrowed the number of chemicals needed from in excess of 300 to possibly 12. Several chemicals may increase activity of the cyclooxygenase and lipoxygenase pathways; consequently, a search was made to identify one or two chemicals which most effectively block each of those pathways. Control of calcium, chloride, potassium, sodium, and other electrolyte

Neutralizing Chemicals

FOODS	Apiol	Caffeine	Camphor	Capsaicin	Cinnamaldehyde	Cinnamic Acid	Coumarin/Scopoletin	Eugenol	Folic Acid	Gallic Acid	Indole	Isoascorbic Acid	Malvin (Anthocyanidins)	Menadione (Napthoquinones)	Niacin	Nicotine	Phenylalanine	Phenyl (Benzyl) Isothiocyanate	Phlorizin or Phloridzin	Piperine	Piperonal	Riboflavin	Rutin-Quercetin (Flavonol)	Saccharin	Thymine	Thymol	Vanillylamine
Allspice										●																	
Almond	●							●		●											●						●
Anise	●							●																			
Apple						●	●			●			●	●					●				●				●
Apricot						●				●											●		●				●
Avacado						●				●			●	●									●				●
Banana						●	●			●			●	●				●			●		●				●
Barley							●			●			●										●				
Bean, kidney													●					●		●			●				
Bean, string													●					●		●			●				
Beef	●						●	●								●		●	●	●	●		●				●
Beet										●	●			●							●		●				
Beet sugar							●			●									●	●	●		●				
BHA/BHT										●																	
Blueberry										●			●														
Blue food color*										●																	
Brocolli																		●			●						
Buckwheat	●																						●				●
Cabbage										●			●					●			●						
Calif. Bay Laurel	●							●																			
Cane sugar							●			●													●				
Cantaloupe										●											●						●
Carrot	●						●	●		●			●										●				
Cashew										●						●				●			●				●
Celery	●					●	●	●		●			●	●									●				
Cheese	●					●	●	●		●								●	●				●				●
Cherry										●			●	●							●		●				●
Chicken							●						●					●		●			●				
Cinnamon					●	●	●			●											●		●				●
Cloves	●					●		●		●			●														●
Cocoa/chocolate	●	●					●	●		●				●				●					●				●
Corn							●			●			●										●				
Cream of Tartar										●																	
Cucumber										●										●	●						
Dill	●							●																			
Egg							●			●			●	●				●		●			●				
Garlic																		●		●							●
Gelatin																			●				●				●
Ginger										●																	
Grape						●				●			●	●					●	●			●				●
Halibut						●				●			●						●	●							
Honey, clover										●										●	●						
Honey dew melon										●																	
Lamb							●						●					●	●	●	●		●				●
Lemon	●						●	●		●													●				●
Lettuce						●	●			●									●		●		●			●	●
Lime										●																	
Mace	●							●																			

Figure 15.1 Chemicals found by R. W. Gardner to neutralize reactions to specific foods.

Neutralizing Chemicals

FOODS	Apiol	Caffeine	Camphor	Capsaicin	Cinnamaldehyde	Cinnamic Acid	Coumarin/Scopoletin	Eugenol	Folic Acid	Gallic Acid	Indole	Isoascorbic Acid	Malvin (Anthocyanidins)	Menadione (Napthoquinones)	Niacin	Nicotine	Phenylalanine	Phenyl (Benzyl) Isothiocyanate	Phlorizin or Phloridzin	Piperine	Piperonal	Riboflavin	Rutin-Quercetin (Flavonol)	Saccharin	Thymine	Thymol	Vanillylamine
Malt																				●							●
Mango										●																	
Maple syrup										●																	
Milk (cow)	●					●		●		●			●					●	●	●	●		●				●
Milk (human)																●											●
Mints																										●	
Mustard																		●			●						
Nectarine																					●						
Nutmeg	●							●		●			●							●							●
Okra										●											●						
Onion						●				●			●	●				●		●			●				●
Orange	●					●	●	●		●									●				●				●
Parsnips						●				●									●				●				
Peach										●											●						●
Pear										●				●						●			●				
Pea	●					●	●	●		●	●			●				●		●			●				
Peanut																		●		●							●
Pecan										●			●														●
Pepper, black	●					●		●		●								●		●	●		●				
Pineapple						●				●			●								●		●				
Pork																				●			●				
Potato							●			●						●				●			●				●
Pumpkin										●											●						
Rabbit																				●							
Radish										●			●					●									
Raspberry										●			●	●									●				
Red food color*										●																	
Rice							●																●				
Sage																										●	
Salmon													●						●	●			●				
Sassafras	●							●																			
Soybean	●						●	●					●					●	●	●			●				●
Spinach										●																	
Squash										●				●							●		●				
Strawberry						●					●		●	●									●				●
Sweet potato																			●				●				
Tapioca						●	●																●				
Thyme																										●	
Tomato	●					●	●	●		●	●		●			●		●		●	●		●				●
Trout						●													●	●			●				
Tuna							●												●	●							
Turkey							●													●							
Vanillin/vanilla										●										●			●				●
Venison																			●	●	●						●
Walnut	●							●		●			●														
Watermelon										●			●								●		●				
Wheat							●			●													●				
Yam										●									●				●				
Yeast							●									●			●	●	●		●				●

Figure 15.1 (continued)

channels is likewise valuable in therapy inasmuch as these chemicals may be the activating forces in initiating adverse effects, as outlined in Chapter 1. I have found that benzenesulfonic acid (sodium salt) has been a key chemical in arresting and/or preventing reactions since it theoretically acts as a calcium entry blocker, much as verapamil. Magnesium is also effective in blocking the effects of calcium, but must be used at correct dosage levels since this mineral may also cause side effects. Both benzenesulfonic acid and verapamil at neutralizing doses will arrest reactions to calcium chloride and norepinephrine. Taurine (2-aminoethanesulfonic acid) was discussed in an earlier chapter as a calcium channel modulator. The sulfonic acid attached to the benzene ring appears to be even more effective in controlling calcium channels than when attached to a linear molecule such as in taurine. Benzenesulfonic acid (sodium salt) could conceivably be acting as an inhibitor of phospholipase A_2 activity as well. Evidence of this relationship came by activating a reaction using sublingual challenge with porcine phospholipase A_2 and then finding that the neutralizing dose of benzenesulfonic acid effectively aborted the reaction. Quercetin has been reported to inhibit this enzyme, which is actively associated with the release of arachidonic acid from triglyceride.[1] I verified this effect using a neutralizing dose of quercetin. Propyl gallate was also effective in arresting the response to this lipase.

Benzyl isothiocyanate, which is present in substantial amounts in mustard, onion, garlic, cabbage, and the legume family, has been a major cause of stress, particularly gastrointestinal. I initially found control of this reaction centered around a neutralizing dose of either esculetin (6,7-dihydroxycoumarin), α-tocopherol, glutathione, cysteine, or selenium. Glutathione and α-tocopherol have been the chemicals of choice because of their ability to neutralize reactions to butylated hydroxyanethole (BHA) and butylated hydroxytoluene (BHT), to decrease calcium release from platelet membranes, to inhibit release of serotonin as well as thromboxane, and to interfere with the phospholipase enzyme release of arachidonic acid. More recently, I have found that indomethacin, vanillylamine, benzyl imidizole, and lithium benzoate are effective in controlling reactions to the chemicals listed above, as well as reactions to uric acid, urea, xanthine, histamine, acetylcholine, solanidine, choline, serotonin, tryptamine, melatonin, indoleacetic acid, arachidonic acid, sodium hypochlorite, sodium sulfite, oxytocin, ethanol, and salsolinol. Since imidazole and benzyl imidazole have been identified as thromboxane synthase inhibitors,[2,3] one might surmise that the common effect of some of these chemicals is associated with thromboxanes. Some common denominators have to be found. The sulfur compound effects on acetylcholine receptors and the interaction of lithium ions with both acetylcholine and serotonin offer clues, but what is the common denominator?

After I experienced a convulsive seizure while sleeping, my medical doctor prescribed phenobarbital in an effort to prevent any further problems. I reacted to the medication and consequently studied the possible pharmacological reason. This medication is given to increase the efficacy of γ-aminobutyric acid (GABA) synapses. GABA acts as an inhibitory neurotransmitter in the mammalian central nervous system. Testing disclosed that I reacted to GABA as well, so I prepared a neutralizing dose of GABA and that dosage erased a reaction to phenobarbital also.

Ethyl alcohol (ethanol) has been needed as a solvent for some of the chemicals which are insoluble in water or alkaline solutions. When testing with chemicals in solution in ethanol a judgment had to be made as to the real cause of a reaction: was it due to the ethanol or the chemical being tested? The finding that ethanol is converted to salsolinol, a phenolic compound, led to the discovery that the neutralizing dose of salsolinol would arrest the physiological effects of ethanol (described in Chapter 9). Accordingly, if a reaction was observed, the neutralizing dose of salsolinol was tested first to eliminate that possible cause. An alternative finding has been that lithium will blunt the mood change associated with ingestion of alcohol and may provide useful therapy for alcoholics, provided it is used at neutralizing dose levels. Vanillyamine can also be substituted for salsolinol or lithium.

Gallic acid, as mentioned above, was found to effectively stop reactions to many foods (see dot chart — Figure 15.1). This phenolic exhibited the same cyclooxygenase inhibiting action as aspirin since it could be interchanged with aspirin or indomethacin in curbing reactions to several chemicals in common. Gallic acid, or propyl gallate, was found to be specific for a reaction to naphthoquinones such as vitamin K (menadione, for example). Neither aspirin nor indomethacin was useful in controlling that reaction.

Quercetin appears to have lipoxygenase inhibiting potential inasmuch as this flavonol is effective in inhibiting reactions to aspirin, gallic acid, and other cyclooxygenase inhibitors. Cinnamic acid is shown in the dot chart. Subsequent study led to caffeic acid (3,4-dihydroxycinnamic acid), a more active form of cinnamic acid and one of the most potent inhibitors of 5-lipoxygenase, due in part to its catechol structure. Caffeic acid is formed upon hydrolysis of chlorogenic acid, which is widespread in plants (including fruits).

Coumarins were also found to be lipoxygenase inhibitors. Esculetin (6,7-dihydroxycoumarin) and umbelliferone (7-hydroxycoumarin) were valuable in countering inflammatory reactions to several foods and inhalants; esculetin was the most effective, but it is also very expensive.

Quantities (dosage levels) of the chemicals were critical in either provoking or arresting reactions. A 1:10 dilution series, starting with 1% stock solutions, was used initially. Then the 0.005 *M* solution series was found to be more practical (see Chapter 14). An example of preparation of a 0.005 *M* solution of gallic acid follows. The molecular weight of gallic acid is 170.12 g. A 0.005 *M* solution would be $0.005 \times 170.12 = 0.8506$ g/l. If only 100 ml are needed, the figure becomes 0.085 g. For convenience, I suggest a 1% stock solution of the compounds and then the use of a pipette to measure the quantity needed. In this case, the calculation becomes 0.085/0.01 = 8.5 ml + 91.5 ml solvent (distilled water for this chemical). This then provides 100 ml of a 0.005 *M* solution. Three dropper bottles are needed for each chemical. They include the concentrate (0.005 *M*), a 1:1 dilution of the concentrate, and a 1:9 dilution of the concentrate. The procedure of application is initiated by applying one drop of the 1:9 dilution sublingually, then waiting 10 to 15 seconds, apply one drop of the 1:1 solution, wait 10 to 15 seconds, and then finally apply three drops of the 0.005 *M* solution.

Application of 3 drops (0.15 ml) of the 0.005 *M* solution of gallic acid sublingually is quantitatively equivalent to approximately 130 µg. This is an infinitely small amount, but it emphasizes the degree of sensitivity of body systems to these chemicals. Repeated application of the chemicals over months or years results in a need for increased amounts to neutralize reactions. For instance, I moved from the 0.005 *M* concentration to the 1.0% solution, doubled that to a 2.0% solution, which was doubled again to 4%; I then finally started incorporating the chemical powders in no. 2 and no. 1 gelatin capsules. Doing so has moved the dosage from microgram levels to 100 to 150 mg. At the higher levels protection is extended 4 to 5 hours. Delayed release tablets would offer still another advantage, particularly during sleep.

A patient must be symptom-free in order to demonstrate an active response to chemicals by the sublingual method. Then a drop of the 1:1 solution can be used (without use of the 1:9 dilution first) to provoke symptoms. Patients presenting with symptoms should be given the 1:9, 1:1, 0.005 *M* sequence to determine if the chemicals in use will relieve their ongoing reactions. Normally the 1:9 dilution will neutralize reactions, but several chemicals may be required before there is a complete remission of symptoms. A reverse order of "challenge" testing of those same chemicals will identify which of those used is actually essential for the patient. An increased pulse rate from my normal pulse of 48 to 50 beats/minute to 60 to 80 beats/minute has accompanied most of my reactions, particularly upon sublingual challenge using non-neutralizing doses of the chemicals. Respiratory changes, cerebral effects, etc., may be more prominent in other patients.

Disturbed sleep during the rapid-eye-movement periods of sleep has awakened me usually two times each night due to nocturnal penile erections. A solution to that problem has taken months because of difficulty in finding the chemical which most effectively controls the reactions to uric acid, testosterone, 5-hydroxyindole-3-acetic acid, and melatonin (a derivative of serotonin formed during the rapid-eye-movement of sleep). Indomethacin (a cyclooxygenase inhibitor) was tagged as a chemical which would arrest the reactions to serotonin and melatonin (Figure 15.2). This suggests that PGE_2, PGI_2, and other vasodilating PGs are partially responsible for the abnormal tubenescence. Aspirin was needed in addition to

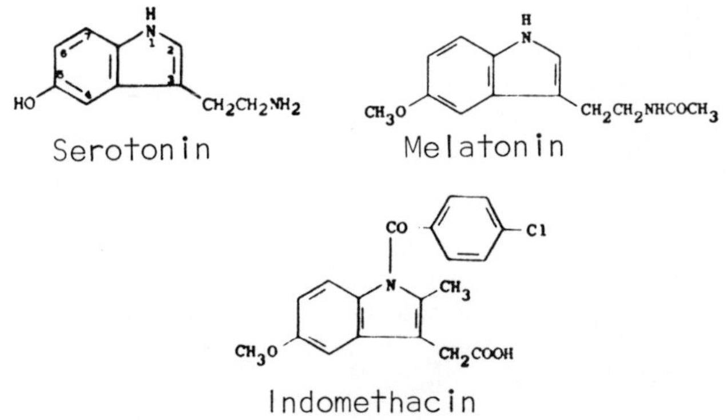

Serotonin

Melatonin

Indomethacin

Figure 15.2 Comparative chemical structures of serotonin, melatonin, and the anti-inflammatory agent indomethacin.

indomethacin, leading to more evidence of the role of PGs in creating this vasodilitation. The finding that uric acid and testosterone also were causing reactions and that these reactions could be controlled by gallic acid and caffeic acid points to a possible dual activation of this problem by products of both the cyclooxygenase and lipoxygenase pathways. Another observation: lithium benzoate (but not chloride or carbonate) neutralizes a reaction to melatonin.

A recent discovery was that lithium carbonate was needed to arrest a reaction to inositol. A clue to this effect came from a most informative paper by Berridge.[4] He points out that lithium has a specific action of inhibiting the recycling of inositol which is normally converted to 1,4,5-triphosphate, a compound which functions as an intracellular calcium-mobilizing second messenger. He makes the additional observation that lithium is a homeostatic drug which seems to be selective against hyperactive receptors, driving them back toward their normal operating range. Lithium should only be administered at neutralizing dosage levels, as has been indicated for the other chemicals.

CONCLUSIONS

In my experience, after many years of study, I have found neutralizing doses of the following chemicals to be most effective in controlling reactions to foods, pollens, and chemical exposures: quercetin, caffeic acid, gallic acid, benzenesulfonic acid (Na salt), indomethacin, aspirin, glutathione, esculetin, benzyl imidazole (or imidizole), τ-aminobutyric acid (GABA), lithium benzoate, zinc sulfate, vanillin, and octopamine. The chemicals I recommend to medical doctors for use on an initial trial basis include benzenesulfonic acid (Na salt), quercetin, gallic acid, caffeic acid, indomethacin, aspirin, vanillin, glutathione, nicotinamide, lithium benzoate, and zinc sulfate.

REFERENCES

1. **Lee, T. P., Matteliano, M. L., and Middleton, E., Jr.,** Effect of quercetin on human polymorphonuclear leukocyte lysosomal enzyme release and phospholipid metabolism, *Life Sci.,* 31, 2765, 1982.
2. **Whittle, J. R., Kauffman, G. L., and Moncada, S.,** Vasoconstriction with thromboxane A2 induces ulceration of the gastric mucosa, *Nature,* 292, 472, 1981.
3. **Walker, K. A. M.,** Selective inhibitors of thromboxane synthetase, in *Handbook of Eicosanoids: Prostaglandins and Related Lipids, Vol. 2, Drugs Acting via the Eicosanoids,* Willis, A. L., Ed., CRC Press, Boca Raton, FL, 1989, 165.
4. **Berridge, M. J.,** Inositol triphosphate, calcium, lithium and cell signaling, *J.A.M.A.,* 262, 1834, 1989.

APPENDIX

Use of Phenylated Food Compounds in Diagnosis and Treatment of 100 Patients with Food Allergy and Phenol Intolerance*

I. CHARACTERISTICS OF PATIENT POPULATION

1. Age, sex, duration of illness
   a. Average age, 32
   b. 80% females
   c. Average duration, 7 years
2. Clinical severity classification

		% Pts.
a. Class I:	Moderately symptomatic; able to work; many nonreactive foods.	0%
b. Class II:	Moderately symptomatic upon petrochemical exposure; able to work in controlled environment using charcoal mask and air filters at home, work, and in auto and with open windows in office; only 20 to 40 nonreactive foods; pesticide-free foods required on 4-day rotation basis.	40%
c. Class III:	Severely symptomatic — foods and chemicals; unable to work even in environment controlled with mask and filter; at home; only 10 to 20 nonreactive foods; pesticide-free foods required on 4 to 7 day rotation basis; natural fiber clothing only.	55%
d. Class IV:	Totally disabled; only 0 to 4 nonreactive foods; total environmental control required, including	5%

* From McGovern, J. J., Jr., Gardner, R. W., and Brenneman, L. D. Paper presented at the 37th Annual Congress of the American College of Allergists, Washington, D.C., April 4-8, 1981.

protective isolation in desert, mountain, or oceanside (e.g., unable to travel without oxygen).

3. Physical examination — abnormal findings
   a. Present in 100% of patients
      1) Marked facial pallor, without anemia
      2) Moderate edema of margin of both upper eyelids
   b. Most frequent abnormalities on examination
      1) CNS (central nervous system): alterations in level of alertness and comprehension during examination
      2) Muscle: tender palpable spasm, muscle groups of back and neck
      3) Respiratory: <u>hypo</u>ventilation, cyanosis
      4) Gastrointestinal: visibly distended abdomen
      5) Ear, nose, and throat: retracted TM (retracted tympanic membrane)
      6) Skin: petechial hemorrhages and ecchymoses
      7) Joints: swollen finger joints
      8) ANS: alternating cold and hot; dry, blue, white, red hand(s)
   c. Objective findings in multiple systems
      1) CNS plus one other organ system 26%
      2) CNS plus two other organ systems 35%
      3) CNS plus three other organ systems 24%
      4) CNS plus four other organ systems 15%
4. Previous hospitalization
   a. 44% hospitalized at Environmental Control Units (Dallas, Chicago, Denver)
   b. Hospitalized 3 months to 3 years prior to present study
5. Important common clinical factors
   a. Intractable pain syndromes in all cases of present study following ingestion of certain foods: head, face, neck, gastrointestinal tract, musculoskeletal system
   b. Frequent, adverse reactions to medication, especially medication containing the phenol molecule
   c. Cognitive disruption following food ingestion
      1) Decreased concentration, memory and mental acuity
      2) CNS depression: withdrawal — depressive reaction with suicidal urges
      3) CNS elevation: agitation — hyperkinetic reaction with violent impulses
   d. Heightened perceptual awareness and adverse response to olfactory stimuli of petrochemical origin
      1) Aromatic: cosmetics, pesticides

2) Combustion products of petroleum: natural gas, auto exhaust

e. Diminished adaptive capacity to petrochemicals
   1) Gradual — chronic, low-level exposure (months to years)
      a) Phenol: allergists, photographers
      b) Formaldehyde: clerical workers in energy-efficient buildings
      c) Hydrocarbons: refinery workers, organic chemists
   2) Abrupt — brief, high-level exposure (hours)
      a) Toxic spills (e.g., MEK)
      b) Iso thio (benzyl isothiocyanates)
      c) Fugitive fumes

6. Laboratory abnormalities
   a. Disorder of the immune mechanism, consistent with immunotoxic reaction
   b. Abnormalities observed — alone or in combination
      1) Elevated circulating immune complexes
      2) Elevated prostaglandin $F_{2\alpha}$
      3) Decreased complement
      4) Decreased IgE; RAST usually negative
      5) Decreased IgA
      6) Elevated IgM
      7) Decreased T and elevated B cells

7. Biologic evidence for inadvertent modification of host defense mechanism, secondary to immunotoxic reaction
   a. Animal model for phenol
   b. Animal model for chlordane
   c. Animal model for organophosphate pesticides

8. Laboratory blood tests
   a. Circulating immune complexes .....36% abnormal elevated values
   b. Prostaglandin $F_{2\alpha}$ ....44% abnormal elevated values
   c. Complement C_3/C_4 ....55% abnormal depressed values
   d. T-Cell assay (Normal range = 0.9 to 2.9 m^3)

	Absolute total T-cells/mm^3							
	Below normal range	Percentile of range						
		.10	.20	.30	.40	.50	.60	.70
% of patients	45	10	25	5	10			

Note: This data indicates that most patients had T-cell values in the lower end of the normal range or had frankly depressed values.

e. B-Cell assay (Normal range = 8 to 10%)

	Range of values		
	Below normal	Normal	Above normal
% of patients:	10	24	66

f. IgE assay

	I.U.							
	0-10	11-20	21-50	51-100	101-200	201-300	301-400	701-1000
% of patients:	15	15	25	0	10	20	0	10

g. RAST test (20 inhalant antigens tested)

	% Positive tests				
	Less than 10	11-25	26-50	51-89	90-100
% of patients:	38	20	14	12	14

9. Intradermal allergy skin test protocol for each patient
   a. Serial dilution titration method of provocative challenge was used.
   b. An average of 30 individual antigens were tested, intradermally:
      1) 3 representative tree pollens
      2) 3 representative grass pollens
      3) 3 representative weed pollens
      4) 3 to 4 common molds
      5) Cat, dust, feathers
      6) Ethanol, histamine
      7) 8 to 10 major foods
      8) 12 phenyl food compounds (sublingually)
   c. Positive skin test reaction was determined by expanded wheal growth; positive, systemic reactions were determined by measured, objective systemic findings following injection.
   d. Initial challenge dose in each case:
      1) 0.1 mg, all inhalant extracts, food } intradermal
      2) 1.0 mg, ethanol } intradermal
      3) 0.2 μg, histamine } intradermal
      4) 1×10^{-5} mg phenyl food compounds — sublingual

II. OFFICE TREATMENT OF SYSTEMIC REACTION PROVOKED BY INTRADERMAL INJECTION AND ASSOCIATED WITH POSITIVE WHEAL GROWTH

1. Serial injection of progressively smaller doses of antigen at 10-minute intervals until negative wheal growth occurred and signs and symptoms abated. (*C. Lee and J. Miller*)
2. When faced with continued systemic signs and symptoms in presence of negative wheal growth, we repeated injection of same dose, in mg, that gave negative wheal growth. (*P. Peters*)
3. Patient response to office treatment of antigen-induced systemic reaction:

% of patients		% of symptom relief
18	*achieved*	100
21	*achieved*	95
29	*achieved*	90
11	*achieved*	85
11	*achieved*	80
3	*achieved*	75
2	*achieved*	70
1	*achieved*	65
2	*achieved*	60
1	*achieved*	55
1	*achieved*	50

4. When we were faced with severe, systemic reaction from the first dose of a phenyl food compound, given sublingually, which continued for more than 10 minutes, we gave another dose of 1/10th to 1/5th as much of the same antigen if the patient had sustained a pulse response of more than 10 bpm within the first 3 minutes after the first dose had been given. 15% of our patients responded with tachycardia to phenyl food compound testing. In each case, we repeatedly gave progressively smaller doses until we reached a dose of antigen which produced no pulse elevation. Complete suppression of symptoms almost always occurred at the dose level that caused no pulse rise.

When a systemic reaction occurred from the first test dose of a phenyl food compound antigen in the absence of any change in the pulse rate, we found that the further administration of a 5-fold or 10-fold incremental increase of the antigen which produced the signs and symptoms often decreased the severity of the reaction. When that happened, we found that we could administer another 5- or 10-fold incremental increase of the antigen and often obtain suppression of the signs and symptoms brought on by the first dose.

From the analysis of 700 such reactions on 100 patients over a 12-month period, we came to realize that most patients respond to these phenyl food compounds with a nonmonotonic dose-response curve. In a given patient, the intensity of the reaction is much greater and the rapidity of onset is much quicker than the response to a common food extract, although the clinical characteristics may be the same in both cases.

Index

K

L

M

N

Q

R